CW01497195

A HISTORY OF THE COTTON INDUSTRY

It may assist us to form a conception of the immense extent of the British cotton manufacture when it is stated, that the yarn spun in this country would, in a single thread, pass round the globe's circumference 203,775 times; it would reach 51 times from the earth to the sun, and it would encircle the earth's orbit eight and a half times. To complete the wonder – the manufacture is the creation of the genius of a few humble mechanics; it has sprung up from insignificance within little more than half a century.

Edward Baines,
History of the Cotton Manufacture in Great Britain, 1835

Let any great social or physical convulsion visit the United States, and England would feel the shock from Land's End to John O'Groats. The lives of nearly two million of our countrymen are dependent upon the cotton crop of America. Should any dire calamity befall the land of cotton, a thousand of our merchant ships would rot idly in dock; ten thousand mills must stop their busy looms; two thousand mouths would starve for lack of food to feed them.

E.N. Elliott,
Cotton is King and Pro-Slavery Arguments, 1860

A HISTORY OF THE
COTTON
INDUSTRY

A STORY IN THREE CONTINENTS

ANTHONY BURTON

PEN & SWORD
TRANSPORT

AN IMPRINT OF PEN & SWORD BOOKS LTD.
YORKSHIRE - PHILADELPHIA

First published in Great Britain in 1984 by the BBC under the title, The Rise and Fall of King Cotton. Republished in this format in Great Britain in 2023 by Pen and Sword Transport
An imprint of
Pen & Sword Books Ltd.
Yorkshire - Philadelphia

ISBN 978 1 39905 731 8

Typeset in 10/12.5 pt Palatino
Typeset by SJmagic DESIGN SERVICES, India.
Printed and bound in the UK by CPI Group (UK) Ltd., Croydon. CR0 4YY.

Pen & Sword Books Ltd. incorporates the imprints of Pen & Sword Books:
After the Battle, Archaeology, Atlas, Aviation, Battleground, Discovery, Family History, History, Maritime, Military, Naval, Politics, Railways, Select, Transport, True Crime, Fiction, Frontline Books, Leo Cooper, Praetorian Press, Seaforth Publishing, Wharncliffe and White Owl.

For a complete list of Pen & Sword titles please contact

PEN & SWORD BOOKS LIMITED
George House, Units 12 & 13, Beevor Street, Off Pontefract Road, Barnsley, South Yorkshire, S71 1HN, England
E-mail: enquiries@pen-and-sword.co.uk
Website: www.pen-and-sword.co.uk

or

PEN AND SWORD BOOKS
1950 Lawrence Rd, Havertown, PA 19083, USA
E-mail: uspen-and-sword@casematepublishers.com
Website: www.penandswordbooks.com

CONTENTS

PREFACE TO THE FIRST EDITION

The process of industrialisation is central to our whole modern civilisation. In this book, I have looked at the growth of one industry, cotton textiles, in terms of three societies. The first is India, where the use of cotton can be traced back for millennia – a society where the intervention of European traders seemed at one time likely to destroy the industry forever. That destruction would have been the result of a unique partnership between the other two societies: the American South, which developed as the most important producer of the raw material and Britain, which became the principal manufacturing region. For almost a century, these two depended on each other for their wealth; each developing its own social structures during the period of violent change that we call the industrial revolution. I have tried to explore the processes by which that industrial world was made and the consequences that flowed from its making. Cotton seems to me to be a microcosm in which one can see, in its simplest, most dramatic form, all those forces at work, which have combined to make the world we live in today.

The pattern of development of the cotton industry was complex, involving many centuries and many societies: India was not the only country with an ancient tradition of using cotton: Britain was not alone in developing factories and America never held a monopoly of growing. It is in these countries, however, that the issues can be most clearly seen and it is on these that I have concentrated. This is not, then, a complete history, it is rather a study of change viewed through those societies most affected by it. Change has become the norm of the modern world. The appearance of computers, the silicon chip and the like in recent years suggest that change is going to be with us for a long time. A study of the first, and still the greatest, upheaval of the modern world is not without relevance today.

Inevitably in writing a book of this sort one accumulates many debts; in this case so many as to defy cataloguing. I should, however, like to extend my special gratitude to the staff of the Southern Historical Collection of the University of North Carolina, The Textile Commissioners of Ahmadabad, Bombay and Delhi, the Manchester City Library and, as always, the Bodleian Library, Oxford. And once again, I have to thank the BBC and, in particular, my producer Michael Garrod, for taking up this project for television and extending the scope of the work and Keith Wilton, editor of the series, who for the third time running has managed to make sense of my nonsense. Production of the book was immeasurably helped by the two editors, first Faith Evans and later Sara Menguc, and the book's designer Huw Davies: my thanks to them all.

PREFACE TO THE SECOND EDITION

In preparing this book for the new edition, I have made several revisions to the main text, and added an extra chapter to bring the material up to date. The title has also been changed. The original – The Rise and Fall of King Cotton – was chosen to accompany the TV series of that name. The book has also been redesigned, with the addition of some new illustrations.

Since the first edition was written, several Indian place names have been altered – what was then Bombay is now Mumbai for example. I have kept the original names, when referring to historical periods before the change, and only used the modern version when referring to contemporary places. In recent years there has been, quite rightly, a strong objection to the use of the word 'nigger' in any form. It has only been used here in quotations, because it seemed essential not to dilute the way in which language was used at that time – and sadly not only in America. I hope this is acceptable and will not cause offence.

Chapter 1

TOWARDS A NEW WORLD

A new world was made in the eighteenth century; it was neither a Utopia born in a philosopher's brain, nor a land opened up by exploration. It was our own world, the world of machine and factory. In that period, one country, Great Britain, went through a change so fundamental and traumatic that nothing comparable had happened up to that point in the whole of written history. We call the change the Industrial Revolution, and it is so central to our whole concept of society that it has come to divide the world between what became developed and underdeveloped areas.

The old world was one that was primarily concerned with keeping itself alive. It was an agricultural society, forever balanced on the edge, which marks off comfort from starvation. The new world was dynamic, geared to the notion of continuous economic growth. Very few would argue that starvation is preferable to comfort, nor could they argue that the change from one to the other would have been possible without industrialisation. So, it should logically follow that industrialisation is an essential stage through which a state must pass if its citizens are to enjoy long and decent lives. It does not, however, imply that the process of industrialisation will seem pleasant to those who have to live through the transition. If I have toothache I will feel better after the tooth has been removed, but I do not necessarily enjoy the process of removal. One could say that the industrial revolution was, for many, like having a tooth out without the benefit of anaesthetic. Nowhere can the benefits and the pain be more clearly seen than in the textile industry, and in particular, in cotton. Starting with a few simple techniques, a transformation was begun that was to permanently change the lives of the people of three continents.

Textile making is one of the oldest of man's industries. It is also basically very simple. Take a natural fibre, stretch it and twist it to make a thread. Intertwine the threads and you have cloth. In Britain, the textile industry was founded on wool. It was ideal in an age when transport was both difficult and expensive. The raw material was the fleece of locally reared sheep, and all the processes could be carried out in the same area. Yet, even in the days when wool reigned supreme, other textiles were made. Fibres from the flax plant were used to make linen. And for those who had a taste for, and could afford, something richer, silk could be imported. In seventeenth-century Lancashire, there was a small trade in cotton, imported from the Middle East. The cloth was not considered very grand and was mostly used for lining garments.

A cotton field in Texas.

Domestic textile workers in eighteenth-century Yorkshire in a painting by Julius Caesar Ibbetson. Activities shown include washing the wool, winding and spinning yarn.

Anyone viewing the British textiles scene at the beginning of the eighteenth century would have had an impression of great stability. He might have expected a steady improvement in manufacturing methods, but nothing very dramatic. He would certainly not have looked for any revolutionary change. Why should he? The British woollen trade was the pride of the nation. Travellers noted with delight the wealth it produced. 'It turns', said Celia Fiennes, in *Through England On a Side Saddle* in 1695, viewing the trade in the West Country at the end of the seventeenth century, 'the most money in a week of anything in England.' The poet John Dyer went even further in *The Fleece* in 1757, in verses extolling the happiness and prosperity that filled the manufacturing districts.

Wide around
Hillocks and valley, farm and village, smile,
And ruddy roofs, and chimney-pots appear
Of busy Leeds, up-wafting to the clouds
The incense of thanksgiving: all is joy.

Why should any of this change? No one could prophesy a revolution that would transform the whole basis of the industry.

One of the most astute and careful observers of the British scene was the novelist, essayist and pamphleteer, Daniel Defoe. Between 1724 and 1726, he published his account of a tour through Britain – *A Tour Through the Whole Island of Great Britain*, 1724–6 – in which he gave very full details of the state of the country and especially of its manufacturing districts. Pride of place went to the wool districts of the West of England and more importantly to Yorkshire. Travelling from Blackstone Edge to Halifax, he was astonished to find the wild, hilly country thickly populated. Wherever he looked he saw cloth hung out to dry beside the houses. There was not much sign of life, but when he knocked at the door of one of the master clothiers, 'we presently saw a house full of lusty fellows, some at the dye-vat, some dressing the cloths, some in the loom, some one thing, some another, all hard at work, and fully employed among the manufacture.' Everywhere there was ample evidence that the great woollen industry was thriving.

It is tempting to look back upon such a time as an idyll, a golden Arcadian interlude and certainly, there are aspects of the time whose passing we can mourn. The spinner at the wheel and the weaver at the loom could both work within their own homes and enjoy a certain independence. They were paid for work done, not by the hour, which meant that provided the work was completed, they could choose when to do it. In practice, this meant that many preferred to work long hours in order to enjoy the luxury of time off later. But the picture is, in fact, far more complex. To talk of spinners and weavers is only to tell part of the story.

Wool was bought by merchants who handed it out to the cottage workers. When it came from the fleece it was dirty and greasy, so the first job was to clean

The Jersey wheel: the spinner is drawing out the thread, which is twisted as it flicks off the end of the spindle. A pair of cards can be seen on the floor.

it by soaking in a mixture of urine and water. After drying, the fibres were loosened by heating. Spinning consists of stretching and twisting the fibres, but first they have to be aligned by a process known as 'carding'. The wool was dragged through cards studded with metal wire. It was then ready for spinning on the wheel. It was a slow process and at least five spinners were needed to keep one weaver busy.

After the wheel came the loom. The thread was wound onto a frame known as a warping frame, from which it was fed to the loom as the warp. The mechanism of the loom allows alternate warp threads to be raised and lowered, leaving a gap through which the weft thread can be passed by the shuttle. Warp and weft combine to make the cloth. It is again a simple process, but one that requires enough skill of the weaver to give a man pride in his craft, and, as payment was by the piece, a good weaver could expect more solid reward as well. In good times, the best of them could display their wealth to the world at large, stuffing the money into the bands of their hats. The cloth from the weaver then was sent to be finished.

Here mechanism did come into play in the form of the water-powered fulling mill, where giant wooden hammers pounded the cloth in a mixture of water and fullers earth to shrink and it and remove the grease. The final touch was dressing the cloth, cutting the nap to create a perfectly smooth finish. The cloth dressers were highly skilled and claimed the highest wages. At such times, the textile districts did present something of the cheery appearance noted by Defoe. The women could take their wheels out of doors in good weather and gossip as they worked. Earnings were high enough for weavers to pay homage to their patron saint, Saint Monday, drinking his health right through his name day. In such circumstances, independence was much prized.

The independence was, however, at best only partial. The weavers carried no stock of their own. The merchants supplied the wool and sold the cloth.

They rarely had any money tied up in equipment, so they could ride out any trade slump by simply sitting tight, neither buying nor selling and living on their spare capital. No wool from the merchant meant no work, and spinners and weavers had no spare capital. Such bad times came as often as good.

In view of later developments, it is as well to be aware of the role of the children in the textile families. From an early age, they were put to such simple tasks as carding for the spinners, and soon the boys were expected to take their place in the loom. The latter role was very important, for until the middle of the eighteenth century, the broad loom needed two pairs of hands. It was too wide for one man to cope with the job of throwing the shuttle with its weft from one side to the other. The master took control of passing from one side and also controlling the warp through a foot treadle. The boy's only task was to catch the shuttle and return it, repetitive work that required very little skill. It was usually taken on either by the son of the house or a young apprentice. The work was hard and often tedious, but at least the children worked in their own homes and usually with their own families. That was some compensation.

Trade fluctuated over the years, but the overall work pattern remained stable. Had Britain been an isolated, inward-looking country, then there would have been little reason why things should not have remained that way. But she was not. Britain was a major figure in world trade, exporting her own surpluses and importing exotic goods from other countries. Not only was the country not isolated but, by the beginning of the eighteenth century, was well on her way along the road to colonialism. British colonies had been established in North America and British traders had established their own small enclaves in India in the shape of the East India Company. But they were late arrivals on the Oriental scene. By the time the East India Company had been founded in 1599, the Dutch had already established a near monopoly in the main attraction of the region, the spices of the East Indies. The islands were theirs, and the British had to accept second best in India, and not even that second best was a monopoly, for the French and Portuguese were there as well. The British trade began as a mere fraction of the overall traffic between Europe and the Far East.

The British came to an India that was largely under the control of the Moghul empire. Although the Moghuls never controlled the entire subcontinent, they were by far the most powerful force in the land – and the richest. The capital at Agra showed a degree of opulence scarcely to be matched by any city in the world. It was here that the British traders had to wait, cap in hand, to solicit for trading rights. The ambassadors, men of rank and substance like Sir Thomas Roe, had to follow the court as it moved around the country and had to cope with the capricious nature of the emperor's decrees. Jahangir, for example, seemed less impressed by the news that a handful of English merchantmen had fought off the might of the entire Portuguese fleet to establish trade routes, than he was by the sight of an English mastiff that had been brought as a present, killing a leopard. The truth was that neither Jahangir, nor his successor Shah Jahan, were particularly impressed by European traders nor what they had to offer. But their power was on the wane, and they eventually gave way to pressure and allowed trading posts to be established. Then the hunt was on for goods in India that would find a profitable market back in Britain.

Just as the British had produced textiles from the best available local materials, so too the Indians had made the most of their native crop, cotton. Europeans had known about cotton since at least the fifth century BCE, when Herodotus had described trees which 'Bear fleeces on their fruit, surpassing those of sheep in beauty and excellence'. Now, 2,000 years later, the British began to take an interest. They found many highly sophisticated techniques in use, most of which had been perfected centuries before. The Elder Pliny, writing about 70 CE had described one particular method for decorating cloth. A mordant that would hold a dye was applied, so that those sections after treatment would emerge coloured

The British in India: Mr William Fullerton preparing to greet a visitor.

from the dye vat. In the very best work, the designs were drawn separately and then transferred by laying the pattern on the cloth, pricking it out with fine needles and the rubbing charcoal through the pinholes to create the outline pattern. Then the mordant and dye were added. Using such methods, patterns of great intricacy and beauty could be obtained, and similar effects were later achieved by using delicately carved wooden blocks to print the design. These designs were known as 'chint', later to be Anglicised to the more familiar 'chintz'. Spinning and weaving were combined to produce muslin of exceptional fineness and delicacy. They represented the affluent end of the trade.

Spinning and weaving were universal activities, every village being able to supply cloth to meet its own needs. The craftsmen who produced the best – and most expensive – materials, followed the court on its migrations or set up workshops near the palace of some great noble. The British in their long sojourns at the Moghul court had ample time to inspect such goods, which were so different from anything they had known in Europe. Where the woollens were dull and heavy, these cottons were light and colourful. There was trading potential there,

The choukha: the traditional spinning wheel of India.

but they had set their minds on buying spices. These had the great advantage of being very highly priced for their bulk, making them ideal for long voyages in comparatively small ships. The Dutch still dominated the spice trade, but they had taken a fancy to cotton goods, so a triangular trade was established. Bullion from Britain was used to buy cotton cloth, which was then traded for spices.

Then, in the 1640s, direct trading of cotton goods to Britain began. They were exported from the port of Calicut and given the name 'calico' with various spellings. At first, they proved too exotic for British tastes and the London office issued instructions in 1643 that local designs should be adapted to Western tastes:

> Those which hereafter you shall send me desire may be with more white ground, and the flowers and branches to be in the middle of the quilt as the painter pleases, whereas now the most part of your quilts come with sad red grounds which are not equally sorted to suit all buyers.

By the 1660s, however, the popularity of the new materials was sufficiently well established for patterns to be sent out from England for copying. The Indian craftsmen, however, could make little sense of the designs, so they adapted them to produce something far more exotic and bizarre than their originators intended. What was to become the most popular motif of Indian cloth, the flowering tree, was born from this mixture of styles from two continents. These new materials caused great consternation among the British authorities given the task of assessing the nature of the cloth made from India's strange, sheepish, vegetable, as Samuel Pepys noted in his diary in 1664:

> Sir Martin Noell told us of the dispute between him, as farmer of the additional Duty, and the East India Company, whether callico could be linen or not: which he says it is, having been esteemed so: they say it is made of cotton woole, and grows upon trees, not like flax or hemp. But it was carried by the Company.

But whatever the nature of the cloth, it was soon clear that the strange and exotic was becoming increasingly fashionable and popular. Aristocratic taste was beginning to favour light, easily-cleaned clothing and cotton admirably suited that taste. From being a mere curiosity, the new cloths began to appear as a threat to the British textile industry. Defoe attacked the importation of cotton in *The Trade to India*, (1720), the first of a number of pamphlets he produced on the subject, using his complete armoury of literary weapons, starting with scorn for the fashion:

> The general fansie of the people running upon East India goods to that degree that the chintz and painted calicoes which before were only made use of for carpets, quilts, &c, and to clothe children and ordinary people, become now the dress of our ladies, and such is the power of a mode as we saw our

Part of an eighteenth-century petticoat border of painted and dyed cotton from Madras, with scenes from the domestic life of Europeans in India.

persons of quality dressed in Indian carpets, which but a few years before their chambermaids would have thought too ordinary for them.

He also set out the economic arguments. Money was being taken away from Britain to buy goods abroad, and now the goods the money bought were being sent back home to ruin the local manufacturers. The British were, he declared, 'cutting their Throats with their own Knife'. And if there were any who disagreed with him, they were either fools or knaves:

If this Cause meets with Enemies, if any one Man can be found in Britain, who would not have us leave off Painted Feathers, and stick to our own Manufactures; I say, if one man can be found so prepossess'd, it must be a Man perfectly ignorant in Matters of Trade, and so not worth talking to; or it must be some Callico-Printer, or his Employer, and dependent, who, finding his account in the Mischief, acts upon the corrupt Principle of being willing to get money, tho' at the Expense of the Ruin of his Country.

Like all the best propogandists, Defoe combined his withering attack upon the opposition with a vigorous statement of his own position. It was wool and only wool that had brought and would bring prosperity to the people of Britain. 'Heaven bestow'd the Wool upon them, the Life and Soul, the Original of all their Commerce; he gave it to them, and they have it exclusive of all the Nations in the World; for none come up to it.'

There were very many commentators ready to argue the contrary case – though few who could bring the literary skills of Defoe to the task. John Asgill stated in *A Brief Answer to a Brief State of the Question* in 1719, that it was not the solid, woollen trade that would be affected by the important cottons, but the luxury silk trade; and as the latter was as dependent on foreign yarn as was cotton, there was no reason why the silk trade should be given preferential treatment. The cotton traders had just as good a case if they chose to complain about silk imports. Either way, it would not make the least difference to genuine home-based industry.

The arguments rattled backwards and forwards, and whichever had the better logic, it seemed the protectionists had the louder voice. Excise duty was laid on Indian cotton in 1712 and raised in 1724. There was a total ban on painted calicoes, but plain cloth was allowed in as British printers had established a profitable business of their own. As a Swiss commentator, Jean Rhyiner of Basle noted.

> All the world knows this people, whose industry and plodding patience in overcoming every kind of difficulty, exceeds all imagination. This nation cannot flatter itself with having made many discoveries, but it may glory in having perfected all that has been invented by others.

The criticism had some justice at the time, though it was not to remain so for very much longer. In the meantime, the cotton trade went into a temporary decline and wool remained triumphant. Yet even Defoe, the most vociferous supporter of the wool trade, had to concede that Britain also owed a great deal of her prosperity to overseas trade. She was a maritime nation, and if her ships were to be denied cargoes from the East, then they must be supplied with alternatives. Defoe looked westward across the Atlantic to the North American colonies and, ironically, the trade he proposed was eventually to do more to bring in the hated cotton than anything the East India Company even achieved. But to understand the significance of Defoe's solution we have to look a little more closely at the nature of Britain's colonies, and her policy towards them and look at the place of cotton in the pattern.

Britain was involved at the beginning of the eighteenth century in a complex colonial system, but not one in which a trade in cotton was seen as playing an important part. It was seen as an unwelcome rival to the old well-established woollen industry. In India, cotton was grown, spun and woven as it had been for centuries; it was not a native crop in America. Yet already there were factors at work that were to lead to dramatic changes in the next half century.

Colonial India could scarcely be said to exist at this time and, indeed, India itself had no coherent identity. Rather there were a lot of Indias – India of the decaying Moghul Empire, Hindu India, Muslim India, India of the native princes and in among them a spattering of settlements that made up European India. The Portuguese were obsessed with trying to create a Christian India in Goa, where they were to remain as a colonial power until the second half of the twentieth century. The French and British were far more interested in trade than religion. British policy was clear: the interest of the home country came first. They were there to make profits, not to rule. They looked at the natural resources of the land, the agriculture that included cotton growing and had no interest in getting involved in developing those resources. Even if they had wanted to get involved, they would have had little impact as a small enclave in a vast land. Later, forced into belligerence to protect themselves from various Indian factions, they stumbled into empire. But by then, policy was set along different lines and the will to develop resources was lacking.

Things were very different for the Europeans who went to North America. They found an immense land, thinly populated, with virtually nothing that was valuable for trade. There was apparently none of the gold or silver of South America, but there was an abundance of good agricultural land. The British soon ousted the early settlers, the Dutch, from their colony of New Netherlands and set up a string of settlements along the eastern seaboard. In the north, the settlers found a land rich in game and fish, with good arable land and a climate not too different from that of Britain. It was an obvious lure to the independent-minded, a yeoman prepared to stake out a patch of land, clear it and plant crops. Further south, there was also good agricultural land but with a very different climate, quite foreign to the settlers. It was cold in winter but hot and sultry in summer. The land was often swampy and tropical diseases took their toll. It was potentially rich, but why would you want to work there, when life was far easier further north?

The South may not have been appealing to the yeoman farmer, but it looked very different to the wealthy who could afford to pay others to do the work for them. A plantation system was developed similar to that set up in South America by the Spanish and Portuguese. Virginia, Georgia and the Carolinas were soon busy supplying Europe with its tropical grocery list of rice, sugar and tobacco. But there was still the labour problem to be solved and the solution existed already – slaves. The slave trade to North America had begun with the Dutch in the seventeenth century, and the British soon joined in and, with the Dutch out of the way, secured a monopoly. In 1660, that monopoly was made official with formation of a government approved chartered company, the Royal Africa Company. They were supposed to have a monopoly like the East India Company, but the trade was too lucrative to be contained, and free enterprise traders soon moved in. A triangular route was established: British-made goods went to Africa to be exchanged for slaves, who were then taken to America and American produce brought back to Britain.

The government soon lost interest in trying to protect the Royal monopoly. As far as they were concerned, as long as the ships were British everything was fine, and it was not their responsibility to take an interest in what conditions might be like. The Middle Passage, from Africa to the colonies, was notorious for the suffering of the unfortunate slaves. Eventually, voices would be raised against the horrors of this human trafficking, but they only took effect when the trade had ceased to become profitable. In the early years, there was general indifference, with the Africans regarded as somehow less than human. Even a man as intelligent as Defoe could refer to them in *A Plan of the English Commerce* as 'produce of the British

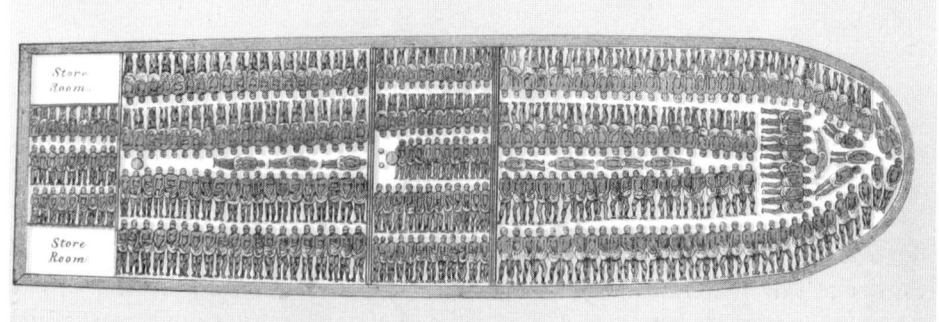

Diagram of a slave ship, from the evidence on the slave trade presented to a British Parliamentary committee, 1790–1.

Commerce'. Thanks to British 'enterprise' this useful commodity could be bought for the equivalent of 30 to 50s in Africa, carried across the Atlantic and sold for £25 to £30. There were other advantages to the trade as well. The money from the sale could be used to buy the produce to take back to Britain, while the planters now had money to buy in luxuries from Europe. Defoe estimated that the trade could be carried on at a rate of 40,000 to 50,000 slaves a year. It was win-win situation for everyone: except of course the Africans.

Slaves were fundamental to the making of the South, fundamental to its very existence, and the system suited the British admirably. It was not merely the trading accounts that appealed. The American South was being established as an area with its own identity – a quiet, acquiescent, semi-feudal society which, in contrast to the independent-minded, self-confident North, was unlikely to cause any trouble. The system seemed ideal – as long as there was a cash crop that could be sold in the European market. The planter could own vast areas of land and huge numbers of slaves, but without that crop it all had little value. As the South grew, so the search for that profitable crop intensified. One potential was cotton, but the big question was – if the South grew cotton would the British buy it? In spite of all the protectionist rhetoric and legislation, the demand for cotton goods persisted in growing. Defoe might deplore it, but he was a realist. The dictates of Parliament cannot always stand firm against the demands of fashion:

> All the Kings and Parliaments that have been or shall be, cannot govern our Fancies. They may make Laws, and shew you the Reason of these Laws for your Good, but two Things among us are too ungovernable, viz, our Passions and our Fashions ... Should I ask the Ladies, whether they should dress by Law, or clothe by Act of Parliament, they would ask me whether they were to be Statute Fools, and to be made Pageants and Pictures of? They claim English liberty, as well as the Men, and as they expect to what they please, so they will wear what they please, and dress how they please.

Chapter 2

THE FLIER AND THE JENNY

There was a growing demand in Britain for the new cotton cloths. There were sources of raw material in the Middle East and India, and a potentially more important source in America. None of this, however, was enough to produce the cloth. There had to be machines and workers to tend them. Britain had the population; her inventors were about to produce the machines.

One of the first essentials for any expansion of the industrial area is that there is enough food available for the workers to be fed. In a primitive society, food production can occupy and feed a very large proportion of the population, with very little surplus. Agriculture in Britain had long since gone past mere subsistence level and productivity was steadily increasing, which meant that fewer workers could produce more food. That does not mean that there was any desire among many of the population to leave the countryside for the town and abandon farm work for industry. A Yorkshire folk song of the period expressed a common view:

I've been forced to work in towns
And here's my litany:
From Hull and Halifax and Hell,
Good Lord deliver me.

Many were indeed forced into the new life. The movement away from small holdings to large estates, with landowners employing wage-earning labourers was well advanced. It was hastened by the spread of enclosures which, while they improved agricultural efficiency, removed the grazing and gleaning rights that had once been enjoyed by the rural poor. The hunt for agricultural efficiency had, as a side effect, produced one of the essentials for a shift to industrial work – a large population needing work.

The total population of the country was subject to fluctuations. In Europe as a whole, a rising swell of numbers had been replaced by a trough, with population declining. It was a familiar story. Increase had brought overcrowding, and that in turn, had brought an increase in disease and a rise in the death rate. Britain, for reasons that are not altogether clear, somehow avoided this trough. There was now a problem that eighteenth-century commentators referred to as a 'surplus population'. Some of the landless poor were sent or made their way to the colonies, which not only eased the problem but created customers. Colonists could

John Kay of Bury, inventor of the flying shuttle.

buy what Britain produced. Those who remained could be set to work making the goods that would meet that demand. It was all as neat and satisfactory as a balance sheet that showed a healthy profit. Men of wealth began to think that their capital might be better employed in investment in trade and industry rather than just buying ever bigger houses. Men with ideas began to think that a lot of money might be made from new inventions. Fortunes were made – but not always, or even often, by the inventors. John Kay was the first in a succession of men whose ideas were to revolutionise industry and who were to see others claim the profits.

Kay came from Bury in Lancashire where, as a child, he was appointed as a reed maker. These reeds were used in the loom. They were split and suspended in the frame and the warp threads passed through. They were then moved by the foot treadles, so that alternate warp threads could be raised and lowered. Kay, being an imaginative man, began to think of ways to improve the loom and started with the parts he knew best – the reeds. He realised that they were fragile and far from ideal and that they could be replaced by wires with a loop in the centre to carry the thread. He set up in business making the new reeds and made a comfortable living. Had he stopped there, his name might have appeared as no more than a footnote in a history of textiles. But he did not. His gaze now turned to the shuttle.

The work of the broad loom weavers was, according to Dyer in *The Fleece*, a thoroughly enjoyable activity for the weaver.

He chuses some companion to his toil.
From side to side, with amicable aim,
Each to the other darts the nimble bolt,
While friendly converse, prompted by the work,
Kindles improvement in the op'ning mind.

Some minds may have been opened, but many more were numbed by the endless repetition of one simple task. More importantly, for an inventor looking for a fortune, it was an expensive waste of manpower. If the shuttle could be kept in a straight line, then there was no reason why mechanical hands could not replace human hands. A suitable shuttle run already existed in the form of the heavy wooden batten, which was used to push the weft threads home after each throw of the shuttle. He put wheels on the shuttle and the shuttle on the batten. Now all he had to do was put little wooden boxes at the ends of the run, each containing a wire 'hand' that could be jerked by a string. The master weaver now only had to pull the handle on the string that dangled in front of him to send the shuttle racing to and fro at a speed far greater than the two-man team had ever achieved. This earned it the name of the 'flying shuttle'. The productivity of the loom could be doubled at very little cost.

Kay took out a patent and sat back ready to make his fortune. He was soon to learn a very unhappy lesson – having the brilliant idea is one thing, profiting from it another. To the clothiers who employed weavers, this was an economic miracle and one much too profitable to share with a reed maker from Bury. Kay's invention was pirated throughout the textile districts. The law, when called on

The flying shuttle in use in Ireland. The weaver is jerking the handle in order to send the shuttle across the loom.

by Kay, very properly fined the offenders when they were brought to court. They responded by getting together to form 'shuttle clubs' to pay the fines of any members who got caught. Pirating was more profitable than paying royalties to the inventor.

The flying shuttle looked less attractive to many weavers, especially to those who had been employed to help the master weaver and who were no longer needed. Throughout the textile districts there were spontaneous outbreaks of machine breaking. Kay's own home was broken into, his loom smashed, and he was lucky to escape unharmed. His hopes of a fortune disappeared under the onslaught. He might have expected the unemployed weavers to be unhappy about his efforts but could reasonably have hoped that those who profited from them might have responded with something more generous than brazen piracy. Disillusioned, he left Britain in the hope of having better luck abroad. He died, disillusioned and poor, in France in 1781.

Kay's invention opened a crack in the doorway to the future, through which men could glimpse a prospect of vast industrial advances. But it was not enough to say that the same number of men could now produce twice as much cloth as before. They could – provided there were twice as many looms, fed by twice as much yarn. Even then there was no point in producing twice as much cloth unless there was a market for it. One obvious option was to make half the weavers redundant and make the same amount of cloth at less cost. There was, however, unlikely to be a huge extra demand for more woollens, but there was an ever-increasing market for cotton goods. Weavers could be employed producing cotton cloth. There was, however, a problem. The flying shuttle had made it possible to weave far more cloth, but there was a bottleneck in the supply of yarn. Where were the extra spinners to come from? With existing spinning wheels, each spinner could only produce one yarn wound onto one spindle. What was clearly needed was a device, which would enable one spinner to produce several yarns. The solution to that problem was to culminate in one of the decisive shifts in the pattern of human life.

The solution might have appeared in 1783, when Lewis Paul took out a patent for a machine for spinning cotton. It could have made his fortune. It could have attracted 100 copyists. It did neither. Paul's career and that of his machine, are surrounded by mysteries. It depended on the principle of passing thread through sets of rollers moving at different speeds, to draw out the raw material in ever thinner threads. The idea was sound as we shall see later, but for reasons we can now only guess at, it failed in practice. Or, to be more precise, it failed commercially. The authorship of the invention is in some doubt. It was claimed by John Wyatt, a Lichfield man and friend of the celebrated Dr Samuel Johnson. Certainly, Wyatt and Paul communicated on the subject of the machine, but John Wyatt firmly claimed it as his own in a letter to Sir Leicester Holt: 'I am the person that was the principal agent in compileing the Engine.'

The most likely explanation is that Wyatt did invent the machine, but being short of funds, allowed Paul to develop and patent it. From that point, at least, the story becomes clearer. Paul was sufficiently confident to install a small-scale version in a building in Birmingham in about 1741. It was circular, with a central drive shaft that turned the various rollers and bobbins were set round the perimeter. Donkeys were used to tun the shaft while two women tended the bobbins, removing and replacing full ones and mending broken threads. It was the first cotton spinning mill. Dr Johnson paid a visit and declared it a success and having bestowed his blessing, left. John Dyer, however, also saw the mill and praised it in verse:

A circular machine of new design
In conic shape, it draws and spins a thread
Without the tedious task of needless hands.
A wheel, invisible, beneath the floor
To every member of th' harmonious whole
Gives necessary motion.

But in spite of the plaudits of literary men, the invention was not a success. A few machines were installed, Wyatt and Paul enjoyed their disputed glory, but then they disappeared from view. Again, the records are vague, but since the machine was practical, the failure may have been down to poor commercial management. The problem of machine spinning had been solved, but the solution had never been generally recognised and for all practical purposes it might just as well never have existed.

As the second half of the eighteenth century got into its stride and the flying shuttle became more widely used, so the demand for a new spinning machine grew. In 1761, the Society of Arts put up a prize for a machine that would 'spin six threads of wool, flax, hemp or cotton at one time, and that will require but one person to work and attend it'. The award certainly acted as an incentive to inventors and numerous proposals were put to the committee. When, however, a completely successful machine was developed, the inventor kept the details to himself, presumably reasoning that he would do better by personally exploiting the machine than he would by handing it over to others. The inventor was James Hargreaves, the machine the spinning jenny.

With Hargreaves we take a large step forward towards the industrial age. As with many new ideas, a mythology has grown up concerning its beginning. There seems to be little truth in the story that Hargreaves had his inspiration when he saw a traditional spinning wheel knocked over on its side. Before he even turned to the problem of spinning, he had already produced a successful machine for use with the earlier process of carding. He was encouraged to continue his researches by a neighbouring farmer and manufacturer, Robert Peel, who was to go on to found a family fortune in cotton, while his grandson was to become prime minster. With the encouragement of a successful entrepreneur and with a successful carding engine already at work, looking at the problem of spinning by machine was the next logical step.

Hargreaves worked at his invention at his home in the tiny Lancashire village of Stanhill, between Accrington and Blackburn. The essence of it was that a horizontal wheel, turned by hand, was used to rotate a number of spindles – at first eight – placed at one end of the frame. The thread was held in a clasp that was pulled back to draw out the thread. Then, in a separate operation, the drawn thread was wound onto the bobbins. The name 'spinning jenny' is no more than a corruption of 'spinning engine' and not a romantic dedication to a former lover. A number of jennies were built and installed in Peel's mill. At first, their use was on a small scale, but word soon got around that there was a new machine that enabled one spinner to do the work of several. To the cottagers, whose livelihoods depended on the spinning wheel, the news brought with it a fear of unemployment. In 1768, Hargreaves' house was attacked, and a score of jennies smashed. Hargreaves made his escape but unlike Kay, instead of leaving the country, he put himself at a safe distance from the Lancashire spinners. He moved to Nottingham, the centre

James Hargreaves' spinning jenny. One operative turning the large handle could spin on sixteen spindles at once.

of the hosiery industry, where there was a steady demand for yarn. He went into partnership in a spinning mill and enjoyed a modest success. Rather belatedly, in 1770, he attempted to take out a patent, but never obtained the protection of the royalties that might have made him a wealthy man.

The jenny was, in fact, accepted quite quickly, for it was soon recognised that it could be used in the home as a sort of super spinning wheel. Others were installed in small workshops where they could be powered by the operatives or by horses and mules. At last, the supply of yarn was beginning to catch up with the capacity of the improved looms, but neither the flying shuttle nor the jenny required any major reorganisation of the traditional industry. They had increased the efficiency of the cottage workers, but the pattern of life remained more or less unchanged. The same could not be said of the next major innovation of the textile industry – the water frame.

Chapter 3

ARKWRIGHT OF CROMFORD

Andrew Ure, a great enthusiast for the new cotton industry, wrote of Richard Arkwright in *The Philosophy of Manufacture* in 1835 as 'a man of Napoleon verve and ambition' who forced through a textile revolution in the face of 'prejudice, passion and envy'. Matthew Boulton, the famous manufacturer of steam engines, accused him of 'tyranny and an improper exercise of power'. Both agreed that in Richard Arkwright the world was seeing a new phenomenon, the factory owner as absolute ruler of his workforce.

Arkwright began his working life as a barber and wig maker, a strange beginning, perhaps, for an inventor of textile machinery, but he did work in Lancashire and the barber's shop is traditionally a great centre for gossip. No doubt he heard his share of the speculation about new spinning machines. In 1761, at the age of 29, he married a girl from Leigh and it was there that he began to work on his designs. In 1768, he persuaded a local clock maker, John Kay – no relation to flying shuttle Kay – to help him produce a working model. They had the use of a room in the Free Grammar School at Preston, and by now Arkwright was so convinced of the ultimate success of his idea that he devoted all his time to

Sir Richard Arkwright by Joseph Wright of Derby. A model of the water frame stands on the table beside him.

perfecting the mechanism. As a result, the family was plunged into poverty and were only rescued when Arkwright persuaded a local innkeeper called Smiley to finance him. That, however, was the year when news of the spinning jenny reached Lancashire and Hargreaves had been forced to escape to Nottingham. When the following year news of Arkwright's work began to spread, he decided to opt for safety first. In 1769 he followed Hargreaves on the road to Nottingham.

In Nottingham, Arkwright met two successful hosiery manufacturers, Jedediah Strutt of Belper and Samuel Need of Derby, both of whom were sufficiently impressed by the model to enter into a partnership. Smiley, who had put up the initial finance was now a junior partner in the enterprise. Strutt, in particular, was receptive to new ideas, for his own fortune was partly built on an improvement he had devised for the hosiery knitting machine that allowed it to be used for ribbed stockings. Arkwright was at last able to give his machine its first full trial. A building was rented and the machinery installed. It consisted, as with Paul's machine, of rollers moving at different speeds to draw out the thread, and rotating spindles to provide the twist. Unlike the Paul machinery, the different parts were set in a vertical frame. Horses were used as the power source. The experiment was a success and now the partners looked for a site where they could begin to work in earnest.

The site they chose was Cromford in Derbyshire, then no more than a tiny hamlet, but it had everything Arkwright wanted. It was tucked away among the

The premises once used by Arkwright as a barber's shop.

hills and its very remoteness was one of its attractions. There were no spinners in the area to feel threatened and there was no one to object to his building a mill and housing for the workers that would turn the hamlet into a town.

Cromford and the mill represented something quite new. The mill itself was, by the standards of the day, enormous. It stood six storeys high, with machinery filling every floor, all powered by one large water wheel. This was more than a mere change of scale from what had gone before, it forced a whole new way of working on the industry. Once the great wheel began to turn and set the rollers turning and the spindles whirring, then the workers had to be in their places. There they had to stay until the great wheel stopped. The factory bell rang out to mark the start of the day's work and rang again in the evening to announce the end. The workers no longer had any choice in how and when they worked; the bell had to be obeyed. With the establishment of the mill at Cromford, the factory age had arrived.

From the first, Cromford attracted a great deal of attention. Wealthy tourists paused in their browsing round the wonders of the Peak District to gawp at the buildings. One visitor, the physician, amateur scientist and man of letters, Erasmus Darwin, celebrated it in verse in *The Botanic Garden* (1791), which manages to explain what is going on in terms of classical allusions:

Where Derwent guides his dusky floods
Through vaulted mountains and a night of woods
The nymph *Gossypia* tread the velvet sod,
And warms with rosy smiles the wat'ry god;
His pond'rous oars to slender spindles turns.
And pours o'er mossy wheels his foaming urns;
With playful charms her hoary lover wins,
And wields his trident while the Monarch spins.
First, with nice eyes, emerging Naiads cull
From leathery pods the vegetable wool.
With wiry teeth revolving cards release
The tangled knots, and smooth the ravell'd fleece.
Slow with soft lips the whirling can acquires
The tender skeins, and wraps in rising spires,
With quicken'd pace successive rollers move,
And these retain, and then extend, the rove,
Then fly the spokes, the rapid axles glow,
While slowly circumvolves the labouring wheel below.

Gossypium is the botanic name for the cotton plant, and all the processes involved in turning the raw cotton into thread are described. The workers removed the cotton lint and then sent it to be carded not by hand but by another new

Part of Arkwright's mill at Cromford. Though the buildings still stand, the original waterwheel seen here has gone.

Arkwright invention, the carding machine. Rollers studded with wire were turned by the water wheel and the cotton passed between them. It was then roughly twisted to form rovings, which were then passed to the rollers of the spinning machine. Because everything was governed by the 'pond'rous oars' the machine became known as the water frame. The mention of glowing axles is no flight of fancy. Friction in the moving parts was considerable and mill fires became commonplace.

The economic importance of the Cromford mill was soon apparent. The whole enterprise was very expensive. For a start, the mill itself cost a great deal, with insurance estimates suggesting it was not less than £3,000. After that, there was the cost of supplying it with water. Then, even after the mill was paid for, a new town had to be built to house the workers who would need to be hired. There was a lot of cash tied up in Cromford, and Arkwright wanted a quick return. He was by all accounts an overbearing and impatient character, but above all else he was ambitious. Here was no inventive genius looking mainly for public acclaim; profit was far more important than praise. He had no intention of sharing the fate of other inventors, and he had made a start by setting up his works in a neutral territory. Well aware of the piratical habits of his fellow industrialists, his first concern was for secrecy. After all, his mechanisms were in themselves quite simple, well within the power of any competent craftsman to copy. That would be easily achieved by just the sort of men for whom he was now advertising in the *Derby Mercury* in December 1771 to make his own machinery for the mill.

An eighteenth-century water frame preserved at the Helmshore Mills textile museum in Lancashire.

Wanted immediately, two Journeymen Clockmakers, or other that understand Tooth and Pinion well: Also a Smith than can forge and file.

Likewise two Wood Turners that have been accustomed to Wheel making, Spoke turning, &c.

That very simplicity made piracy all too easy, so he was soon writing to his partner Strutt:

Desire ward to send those other Locks and allso Some sort of Hangins for the sashes he & you may think best and some good Latches & Catches for the out doors and a few for the inner ons allso and a large Knoker and a Bell to First door.

The preservation of his monopoly was one aspect of Arkwright's search for wealth; his determination to exploit the new machines to the uttermost was another. They had to be kept going as long as possible throughout the working day, and for this to be achieved there had to be a new attitude among the workforce.

Arkwright recruited most of his workers locally, bringing in whole families to occupy the new houses he had built – and built well. One of these in North Street is now owned by the Landmark Trust and can be rented as a holiday home. My wife and two friends stayed in it and found it very comfortable – but it was not quite as it would have been in Arkwright's day. Those who first came there must have been happy to exchange near destitution for work and decent housing. The women and children formed the bulk of the workforce in the spinning mill, while the top floors of the houses were workshops where the men were employed at the loom or the stocking frame. Originally, there had been looms at the mill, but Arkwright soon got rid of them and concentrated on spinning.

The many observers who came to Cromford saw little beyond the miracle of technology. If they watched the women and children at work, they passed no comment. Certainly, few gave any indication that they were seeing a social revolution. Many years were to pass, and Arkwright himself was dead, before the social cost of the miracle was spelt out in any detail. Arkwright's son, also Richard, gave evidence before the Peel committee of 1816, set up by parliament to investigate the conditions under which children worked in the new factories. He described conditions in the early days. The workforce were mostly children, he said, aged from 7 to 13, who worked a thirteen-hour day. He maintained that their health was in no way affected by the work, though he did add, as an apparent afterthought, that some of them had become deformed as a result of the poor design of the machines. The spindles carrying the thread were set very close to the floor and the children whose job it was to change the bobbins and to mend the thread, spent most of their time bent double. Some, it seems, never quite straightened up again. The fault might not, in fact, have been caused by the

Houses for the workers in North Street, Cromford. Note the long windows on the top floors, where the men had their workshops.

machines but could have been the result of malnutrition. In either case, it makes nonsense of the claim that they all enjoyed good health.

Within the confines of Cromford, Arkwright lived the life of a feudal lord. The mill was his castle, as carefully guarded as any fortress, to which only his retainers and a few privileged visitors were admitted. He controlled the lives of his subjects, supplying them with work, home, provisions and every village amenity from church to inn. There are no statistics available for wages in the earliest days, but in the late 1780s, the spinners were earning an average of from 3s 3d to 3s 6d per week. At this time, the mill was working day and night and the night shift earned slightly more, from 3s 11d to 4s 7d. Arkwright received a fortune in profits, the full extent of which became apparent at his death when the estate was estimated to be worth £600,000. He also received the deference of the workforce. He played lord of the manor to the hilt, handing out annual prizes to the butchers, bakers and grocers who, he considered, best served the community. The villagers acknowledged the favours with a hymn of praise to the bountiful master:

Come let us all here join in one,
And thank him for all favours done;
Let's thank him for all favours still
Which he hath done beside the mill.

Modestly drink liquor about,
And see whose health you can find out;

This will I chuse before the rest
Sir Richard Arkwright is the best.

These lines were posted on the inn door at prize-giving and the paternalist received his due. He had been knighted in 1786.

Jedediah Strutt was unlike Arkwright in almost every way. Where the latter was irascible, he was calm-tempered. Arkwright was barely literate while Strutt was not merely literate, but quite a stylist. For a time as a young man, he had lodged with the Wollatt family of Findern and his letters to the daughter of the house, Elizabeth, whom he later married, are positively rhapsodic. 'Ye Findern groves & bowers', he lamented during a temporary absence, 'who haunts your shades now I'm away or hears your warblers sing?' There was also a difference in attitude to business that was far more striking than a mere difference in temperament.

Until the establishment of the mill at Cromford, Arkwright's life had been one of struggle and poverty. In 1771, when business began, he was already nearing his fortieth birthday and as yet had very little to show for his life of work. His drive to ever greater profits was remorseless. Strutt, on the other hand, was already a man of means when he entered into the partnership. He had left farming to establish a hosiery business in Derby, where his ribbed stocking machines were installed. His promotion of machine spinning was at least as much about supplying cheap yarn for his stocking business as it was about profit. The two men had frequent disagreements, and Smalley, the original backer of the scheme, also fell foul of Arkwright's temper. He had been appointed mill manager at Cromford but was soon applying to Strutt for help in easing his quarrels with Arkwright. Strutt could offer little comfort:

I said what I could to persuade him to oblige you in anything that was reasonable & to endeavour to live on good terms at least … you must be sensible when some sort of people set themselves to be perverse it is very difficult to prevent them doing so.

Strutt too must have found Arkwright a difficult partner. In 1778, he built his own mill at Belper, and three years later when Need died, the partnership was dissolved. Strutt was content to build on a local basis, leaving Arkwright to bustle in the world at large. In Belper, Strutt established a similar system to that at Cromford, building houses for the workforce and supplying them with the necessities of life. The work of building town and mills was continued by his son, William, who added the first of a new style of building, a fireproof mill in which the wooden beams and pillars of older mills were replaced by iron. The mill is now home to an excellent textile museum. The houses were, again like Cromford, built to a very high standard. They were in long rows or in square blocks, forerunners of the back-to-back houses which were to predominate in the cotton industry of the nineteenth century. The little cottage hospital, run by Mrs Strutt, provides further evidence that Strutt's paternalism was indeed benign.

The wage books of the Belper mills show very clearly the way in which whole families were involved in the work. The Cotterills, for example, at one time had nine family members in the mill. The father, John, was mainly employed as a labourer, whose average earnings were seldom more than 10s a week. But he was still the main wage-earner. The pay for the rest of the family was on a descending scale, ending with his young daughter, Hannah, who never made more than 2s a week. All told, when they were all at work, the family income was just over £2. Other family incomes followed a similar pattern. The bulk of the work went to the lowly paid women and girls, who outnumbered males in the mill by more than two to one.

There were advantages for working for the Strutts, of which good housing was by no means the least. The rent for excellent houses such as those of Long Row, which still stand, was only in the region of 2s a week The wages quoted, which were for the years 1801–5 were not high, but they compared very favourably with those for agricultural workers at the time. Sir Frederick Eden carried out a personal survey at the end of the eighteenth century – *The State of the Poor* (1797) – and found many families throughout the country sunk deep into the most degrading poverty from which it seemed, none of their efforts could raise them:

> No labourer can at present maintain himself, wife and two children, on his earnings; they have all relief from the parish, either in money, or in corn at reduced price. Before the present war, wheaten bread, and cheese, and about twice a week, meat, were their usual food; it is now barley bread, and no meat: they have, however, of late, made great use of potatoes … Labourers' children, here, are often bound out apprentice at 8 years of age … A very few years ago, labourers thought themselves disgraced by receiving aid from the parish, but this sense of shame is totally extinguished.

These miserable Devon labourers were earning no more than 1s a day, so that they were reduced to begging from the parish to keep themselves and their families

Strutt's North Mill at Belper, dwarfed by the later East Mill behind it.

from actual starvation. And these were not the worst. Many had no work at all and had to rely entirely on inadequate parish relief. If any of them had heard of the promise of work, pay and good housing in Derbyshire, they might have though they were being offered an invitation to paradise.

The new mills were unquestionably improvements over the poor-houses, but, from the earliest days, there were social critics who compared Arkwright's prosperity with the working conditions of those on whose labour that prosperity was built. Arkwright, from being:

> ... a poor man not worth £5, now keeps his carriage and servants, is become the lord of a manor, and has purchased an estate of £20,000; whilst thousands of women, when they can get work, must make a long day to card, spin and reel 5,040 yards of cotton, and for this they have *four-pence or five-pence and no more*.

There was also criticism from the old aristocracy who disliked the prospect of a rising class of men who made their money from trade and manufacture. They scorned the pretensions of the nouveau riche and mocked their possessions. Viscount Torrington viewed Willersley Castle, the hose being built for Arkwright near Cromford:

> Went to see where Sir R.A. is building for himself a grand house in the same castellated style one sees at Clapham, and *really* he has made a *happy* choice of ground, for by sticking it up on an unsafe bank, he contrives to overlook, not see, the beauties of the river, and the surrounding scenery. It is the house of an overseer surveying the works, not of a gentleman wishing for retirement and quiet. But light come, light go, Sir Rd has honourably made his great fortune; and so let him still live in a great cotton mill.

There were grumblings too among the workforce. There was resentment at the more niggling aspects of factory discipline. The complaints were not against long hours and hard work – the poor were used to those – but against the strictness that was seen in all aspects of mill work. Some of the fines in the forfeit book at Belper seem bizarre until you recall that 'the workers' being penalised for such offences as 'Calling thro' the window to the soldiers' or 'Terrifying S. Pearse with her ugly face' were children, perhaps not much more than 8 years old. Some of the more horrifying accounts of the life in these mills come from Samuel Slater who, at the end, of his apprenticeship, emigrated to America, taking the secrets of cotton spinning with him, earning himself the title of 'Father of the American Cotton Industry'. These conditions were tolerated, in spite of grumbles, because the Derbyshire workers had chosen this way of life as an improvement on what they had known before. But success in Derbyshire meant expansion further afield and that, in turn, brought the mills into direct competition with an old, well-established way of life. Mutterings were to give way to shouts of rage.

The success at Cromford encouraged Arkwright to widen his business activities. He began to build other mills in Derbyshire and he also allowed other manufacturers to install his machines under licence, and in some cases entered into partnerships with them. He was beginning to move away from his home base and into the traditional textile heartland of Lancashire when a new mill was begun at Chorley.

One question now remained to be answered – how would the local workers react? Would Lancashire act as peacefully as Derbyshire had done, or would the reactions be as violent as those who had driven out Kay and Hargreaves?

Arkwright had chosen a bad time to open his Lancashire mill. The War of American Independence which had seemed at first to be no more than a local affair in a distant land, involving the suppression of a few colonial malcontents, had, by 1779, grown into a full-scale international conflict. France had joined in on the American side and was attacking British shipping, and the Spaniards were blockading Gibraltar. Inevitably this had an effect on international trade. Work in the textile industry was slack, and workers were in no mood to accept new mills in which one machine could do the work of many.

In October 1779 the Lancashire textile workers rose in their thousands and marched on the mills. The potter, Josiah Wedgwood, visiting the county at the time, wrote this account of events to his partner, Thomas Bentley:

I wrote to my dear friend from Bolton, and mentioned the mob that had assembled in that neighbourhood, but they had not then done much mischief: they only destroyed a small engine or two near Chowbent. We met them on

Arkwright extended his empire after the success at Cromford. This is Masson Mill at Matlock: the central block with the cupola is the original building.

Saturday morning, but I apprehend that what we saw was not the main body, for on the same day in the afternoon a capital engine, or mill, in the manner of Arcrites, and in which he is a partner, near Chorley was attacked, but from its peculiar situation, they could approach it by one passage only, and this circumstance enabled the owner, with the assistance of a few neighbours to repulse the enemy, and preserve the mill for that time. Two of the mob were shot dead upon the spot, one drowned and several wounded. The mob had no fire arms and did not expect so warm a reception. They were greatly exasperated and vowed revenge: accordingly they spent all Sunday, and Monday morning, in collecting fire arms and ammunition and melting their pewter dishes into bullets. They were now joined by the D. of Bridgewater's colliers and others, to the number, we were told, of eight thousand, and marched by beat of drum, and with colors flying to the mill where they met with a repulse on Saturday. They found Sir Richard Clayton guarding the place with 50 Invalids armed, but this handfull were by no means a match for enraged thousands; they (the invalids) therefore contented themselves with looking on, whilst the mob completely destroyed a set of mills valued at £10,000. This was Monday's employment. On Tuesday morning we heard their drum at about two miles distance from Bolton, a little before we left the place, and their professed design was to take Bolton, Manchester, and Stockport in their way to Cromford, and to destroy all the engines, not only in these places, but throughout England.

An army was on the march, storming through Lancashire, leaving behind the ruins of the mills. Arkwright retreated to Cromford and prepared to withstand the siege. The *Manchester Mercury* in October 1779 reported his progress. 'Fifteen hundred Stand of small Arms are already collecting from Derby and the neighbouring Towns, and a great Battery of Cannons ... beside which upwards of 500 Spears are fixt in Poles of between 2 and 3 yards long.' Strutt took similar measure to defend the mills at Belper, and one can still see the gun embrasures cut into the bridge, which runs across the main road from North Mill. They waited for the worst and the worst failed to appear.

The Lancashire army had no real leaders, no carefully formulated plan of campaign. Men who were ready enough to remove the menace from their own doorsteps were less eager to march across the county into Derbyshire, a place as remote to most of them as the coast of Coromandel. They were Lancashire spinners and weavers; let the men of Derbyshire look to their own livelihoods. They had achieved what they had set out to achieve and with Sir George Saville now encamped in the area with three companies of the York Militia, they saw little point in courting further dangers. So, they went back to their homes and the old ways of working. If they looked for any reaction in Derbyshire, they were disappointed. There the mills were no threat to old ways, rather a novelty offering an additional source of employment. The revolt was over.

The Derbyshire mill owners relaxed. Strutt continued the steady development of his Belper base, adding more and grander buildings. For Arkwright, the burning of Chorley mill was a setback, but no more. He set about increasing his empire still further, and then set out to other parts of Britain in search of fresh partners. He dashed furiously around Britain in a coach and four, and on a visit to Scotland in 1783 he met a Glasgow merchant, David Dale. As a result of that meeting, a new mill was built beside the Corra Linn Falls on the Clyde. The site offered many of the advantages that had first attracted Arkwright to Cromford. It had a plentiful and assured supply of water to turn the machinery, and it was remote from any likely arsonists. So, the mill was built, tenements constructed to house the workers and the town was named New Lanark. It was to achieve fame later as the centre of a series of experiments in social organisation by manager, Robert Owen. In *A New View of Society* in 1813, he told the story of how Dale set about finding and training a workforce:

> This however was no light task, for all the regularly trained Scotch peasantry disdained the idea of working early and late, day after day, within cotton mills. Two models then only remained of obtaining these labourers: the one to procure children from the various public charities of the country; and the other, to induce families to settle around the works.

The mills and mill village of New Lanark beside the Clyde.

To accommodate the first, a large house was erected, which ultimately contained about five hundred children, chiefly from work-houses and charities in Edinburgh. These children were to be fed, clothed and educated; and these duties Mr. Dale performed with the unwearied benevolence which it is well known he possessed.

To obtain the second, a village was built, and the houses were let at a low rent to such families as could be induced to accept employment in the mills: but such was the general dislike to that occupation at the time, that, with a few exceptions, only persons destitute of friends, employment, and character, were willing to try the experiment; and of these a sufficient number to supply a constant increase of the manufactory could not be obtained.

Though the business might not have prospered to quite the extent some would have wished, it was a powerful concern, which, together with Arkwright's other ventures, had turned the former barber into a very rich man. But Arkwright was greedy. He pushed through still more patents, even more dubious than the original patent for the water frame. Other manufacturers were not enthusiastic supporters of the idea of handing over yet more royalties. The end of the American Revolutionary War in 1781 had brought a revival of trade, and with that revival there was a cautious return to mill building in Lancashire. It had been a costly business and now they were determined to challenge Arkwright in the courts. In 1785 they challenged the patents in the Court of the King's Bench. John Kay, who had helped Arkwright in the early days and a friend of his, Thomas Highs, claimed authorship of the design of the water frame. Others declared that the machine was no more than an adaptation of Lewis Paul's. At the end of the day, Arkwright found all his patents overturned and he left the proceedings in a high rage, swearing revenge on 'those Manchester rascals'. He threatened to publish the details of all his machines and send them to overseas competitors, but in the end he did nothing and the loss of the patents did little to diminish his increasing prosperity, and when he died in 1792, he was almost certainly the richest manufacturer in the country. The removal of the patents, if it did little harm to his fortune, did wonders for the British cotton industry. The door was wide open for all and sundry, provided of course, they had the capital, and provided they had the cotton to spin.

Chapter 4

FAILURE IN INDIA

By the end of the 1780s, there were just over 100 cotton mills in Britain: fifty years later they had increased more than tenfold and the value of manufacture had gone up from £1 million per annum to over £40 million. And even these dramatic figures tell only part of the story, for while the total value of cotton yarn produced had risen, the cost of the yarn had fallen from 36s per pound to around 3s. This meant an incredible increase by a factor of nearly 500 in the quantity of yarn coming out of British mills, and a corresponding increase in the amount of raw cotton being imported. Estimates vary, but it is generally agreed that by the 1830s, imports of cotton were running at nearly 400 million tons per annum. For cotton to be supplied in these quantities, changes had to be made elsewhere as great as any seen in Great Britain.

There was no shortage of potential suppliers. Egypt, with its long tradition of cotton growing, could provide a small but reliable crop. Plantations had also been established in the New World in Brazil and, more importantly, in the West Indies. The latter grew the native Sea Island cotton, famous for its long staple, creamy-white colour and strong, silky fibres. From the earliest days of settlement on the south-eastern areas of the American mainland, cotton from the West Indies had been considered as a possible crop. An expedition was sent in 1669 by the lords and proprietors of Carolina to establish a settlement that would eventually become Charleston. They received clear instructions.

> Mr. West, God sending you safe to Barbados, you are there to furnish yr self with Cotton Seed, Indigo Seed, Ginger Roots … Yor Cotton & Indigo is to be planted where it may be sheltered from ye North West Winds.

West, who was the leader of the expedition, was later to report back to the proprietors.

> The winter here doth prove something sharpe and colde, soe yt I feare this will not prove a Cotton Country, but our new Commers like it very well, and say they believe it will produce any commodity yt the Charibbe Islands doe, as Cotton, Ginger, Indigo, &c.

In this, the Barbadians were to prove more accurate judges than Joseph West. For a long time, however, the cotton did no more than serve local needs and very little found its way across the Atlantic. Indeed, American cotton was then so rare that

Sea Island cotton. On the right, the cotton is being picked from the tall plant, while on the left the seeds are being removed, using rollers and the cotton is being pressed into bags.

when an American ship landed eight bales at Liverpool in 1784, British Customs officials seized the cargo on the grounds that such a quantity could never have been grown over there. The obvious place to look for steady supplies was still India, the home of cotton. There appeared to be every advantage in looking in that direction, for British influence was growing apace. The few small trading enclaves had expanded vastly to fill the vacuum left by the collapsing Moghul Empire, while the main trading rivals, the French, were virtually finished. European wars had been transplanted to the east, and British forces had come out as victors following the fall of Pondicherry in 1761.

The first years of British rule boded ill for the country. Bengal was taken over and systematically stripped of its wealth, with administrators jostling and pushing each other to see who could acquire the greatest fortune in the shortest possible time. The country was bleeding to death, and the neglected East India Company was hurtling towards bankruptcy. In 1786, Cornwallis was appointed Governor-General and he took the decisive step of separating administration from trading. Men had to choose between the two; the days of using public office for private gain were over. Unfortunately, in solving one problem Cornwallis gave birth to another. In the new division between the trader hoping for a quick profit and the administrator working for a salary, there was nobody in between to encourage local production. So, the planting and growing of cotton was left where it was, just one small part of India's subsistence economy. The small planters grew what they could, sold what they could and paid the new taxes that had been imposed on them. There might still have been an opportunity for the Indian planter to take advantage of the growing demand for their produce, but under British rule a subtle but decisive change had been made in the old pattern of life.

Under the Moghul system, the peasant had paid taxes to a zamindar, who retained a percentage as his own profit as tax collector. It was in everyone's interest to ensure that the peasants did the best they could. It kept them above the starvation line, ensured a profit for the zamindar and provided the revenue for the government. If the zamindar failed to pay up, he was sent for, given a beating and sent back with a warning to do better next time. The beating was passed down the line and everyone hoped for better times to come. A tax collector who failed to deliver the correct amount of taxes was not beaten – that was barbaric. Instead, he was dismissed for incompetence to be replaced by an incorruptible bureaucrat looking steadfastly towards the proper implementation of the law. The peasant, faced by tax demands, had either to meet those demands or account for his actions before a court run by foreigners he did not know, arguing in a language he did not understand. The best he could do was find some way of satisfying these new masters. If the masters required so much cotton, then that was what they would get – and someone else could worry about the quality. But who was to do the worrying? Not the new middlemen. Like the peasants they had to supply so much bulk, and if that bulk was largely made up of below standard material, then that was another problem to be passed on down the line. But the authorities were happy; they had been asked to provide a certain quantity of cotton for Britain, and that quantity had been supplied. The books showed so much cotton had been ordered and that quantity had been delivered. Problems were only discovered when the raw material reached the mills of Britain. And how could anyone trace the faults back through the labyrinthine trading in India? And to make matters worse, even the best cotton was short staple and not well suited to the new machines of the industrial revolution. It had done well enough for the local spinners and weavers, but they too were beginning to suffer under policies decided in far off Europe.

The restrictions on cotton cloth exports to Britain had done nothing to help Indian weavers, and now that the British controlled all the trade, exports were effectively cut to other parts as well. New legislation made a bad situation worse. Taxes were imposed on the movement of cloth between different parts of India, which produced the absurd situation that it was actually cheaper to buy Manchester yarn and cloth than it was to buy the local product. Everywhere, the Indian textile industry was in retreat before the incursions from Lancashire. Not surprisingly, this was deeply resented, as an official of the East India Company pointed out in a letter to the President of the India Board in March 1842 when yet more taxes were added:

> The duty of 10 per cent levelled in this country on the Cotton Goods of India is felt by the natives as a very great grievance. They do not expect that if the duty were reduced, or even abolished, they could compete with the manufacturers of England, but it is felt an unreasonable aggravation of their natural disadvantages, that their hand-manufactured goods should in this country be burdened with a duty of 10 per cent, while the machine-made goods of England, if imported in British ships, are admitted to supersede the manufacture of India on their own soil, at a duty of little more than a third of that amount.

British rule was hampering and hindering the development of all sections of the Indian cotton trade. Lancashire cried out for more cotton. India could have supplied it, if only someone would set the country on the right path.

Looking through the reports of the East India Company, published in 1836, one can chart the progress from optimism to disillusion. In December 1790, the directors gave approval for company ships to carry a two-way trade – convicts out to Australia, cotton back to England. Twenty years later that trade was being described as 'a ruinous and unproductive burthen upon the Company and private importers.' The problems were recognised and analysed:

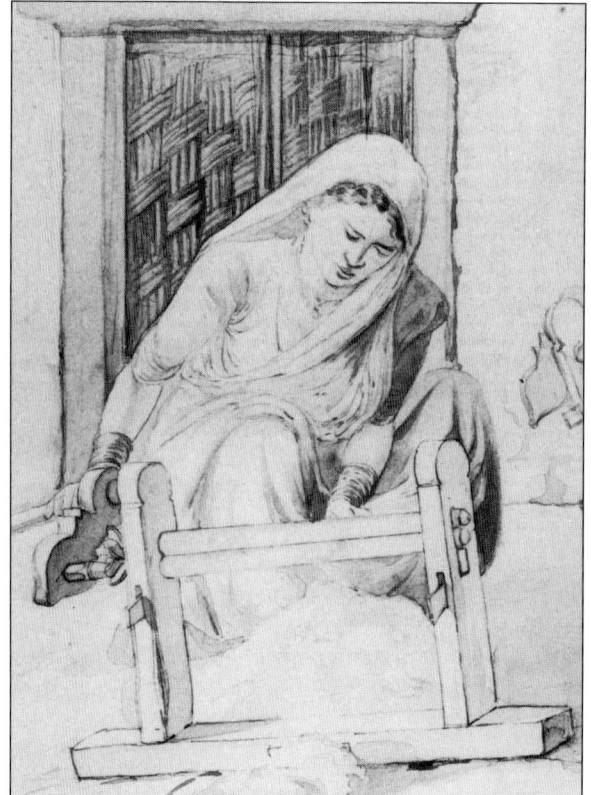

> The cultivators in small farms ... have barely the means of providing for their families and paying their rents; they are incapable of enjoying any

The Indian churkha; the cotton is being passed between the rollers to squeeze out the seeds.

satisfaction from new and successful pursuits, and it would be difficult to persuade them to hazard even the miserable provision they are now certain of, in the hope of obtaining a better one by any new or speculative undertaking.

So, the native Indians could not be expected to speculate in cotton and in the absence of a system that could promise a large return on capital, Europeans were equally loath to put money into production. Methods of cultivation were antiquated, rewards small, and the arguments always came back to the same basic problem. Desperately poor cultivators were at the mercy of the middlemen. Faced with food crops on which they depended for their very existence, and cotton crops on which they hoped to make a tiny profit, the choice was a simple one. Food came first, cotton last – and if some of the plants had rotted in the ground, then there were ways of dealing with that as Thomas Ellison wrote in *A Handbook of the Cotton Trade* in 1858:

> The good Cotton is separated from the seed: and the best stuff, which had been taken away from the good, is beaten with a stone to loosen up the rotten fibre from the seed, and then it is passed through the churkha [rollers used to sperate out the seed]. The good cotton and this bad stuff are both taken into a little room, six feet by six, which is entered by a low door, about eighteen inches by two feet, and a little hole as a ventilator, is made though the outer wall. Two men then go in with a bundle of long, smooth rods in each hand, and a cloth is tied over the mouth and nose; one man places his back so as to stop the little door completely to prevent waste, and they both set to work to whip the cotton with their rods, and to mix the *good and bad together* so thoroughly that a very tolerable article is turned out.

The Court of Directors of the East India Company trumpeted their complaints.

> No excuse will hereafter be admitted by us for the foulness, dirt, and seeds, which are suffered to remain mixed with the cotton: and it is our positive order, that the commissions be not paid to any commercial resident whose provision of cotton shall be faulty in that particular.

They complained but nothing much happened.

Of course, there were ways for improving efficiency, but too often they fell foul of administrative ignorance. A local resident was sent the latest product of advanced technology, a machine for separating the seed from the fibre. Unfortunately, he reported, 'it was not accompanied by any directions, and after cleaning it and carefully examining every part of it, I have failed to discover how it is to be worked.'

Cotton production became caught in a vicious circle. Dirty cotton would not sell in Lancashire and there was no incentive to produce clean cotton in India.

'Awaiting an offer' by Kipling. Indian farmers patiently wait for someone to buy their cotton.

The merchants, who were the ones who would possibly gain from an expanded cotton trade, turned instead to a new crop, on which promised to offer far fewer problems. They set out to satisfy a newly-developed British taste; they turned from cotton to tea.

The story of cotton in India in the nineteenth century is a story of missed opportunities. Yet there was no shortage of men who could see the obvious and express it forcibly and clearly:

> Certainly, without any exaggeration, the most astonishing thing in the history of our rule in India is, that such innumerable volumes should have been written by thousands of the ablest men in the service of the mode of collecting the land revenue, while the question of a thousand times more importance, how to enable the people to pay for it, was literally never touched upon.

Meanwhile, the British looked elsewhere for their main suppliers of cotton. The Indian chance had gone, it seemed, forever.

Chapter 5

THE SLAVE STATES

'The Peculiar Institution' of slavery marked off the Southern American states from their Northern neighbours more absolutely than any geographical or national boundary could ever have done. The differences lay only partly with the morality of slavery; the response of Northerners to the free blacks amongst them, not to mention their treatment of the native North Americans, gave them very little right to claim the high moral ground. What the acceptance of slavery did for the South was to ensure that it followed a totally different pattern of development from that of the North.

The first slaves were brought to North America by the Dutch in the early seventeenth century. There is some doubt about the actual status of the first blacks, for it has been claimed that their position was really that of indentured servants. This legal nicety of definition could have had little meaning for the Africans, carried from their homeland in a stinking hold of a small ship and set down in a foreign land among strangers. In any case, whatever the initial situation, slavery soon became established and legalised in the South – and in some parts of the North as well. But it was in the South that slavery dug the deepest, most pernicious roots. The big plantations formed on the banks of the navigable rivers drew in the slaves and sent out their crops of rice, tobacco and indigo. But the slave was more to the white planter than a mere labourer, he was a symbol of wealth and a marketable asset. And the more slaves came in, the harder it became to envisage a different system.

The plantations in the South fed on slavery. From the earliest days, the South offered a different view of life from that of the North. The European immigrants who had made the long and difficult journey to start a new life saw little to attract them to the South which, with its semi-feudal society, seemed only to offer Europe transplanted. They looked instead to the North, where change and progress seemed to be the watchwords. The poor who might have provided a workforce turned away and the slaves were left to tend the fields. And once that was accepted as slave work, it became virtually impossible to employ white labourers. They could not be brought in to take to take work that set them on the same footing as slaves, and the masters could not, in any case, have employed them. The whole edifice of slavery rested on a theoretical foundation which, proclaimed black inferiority and white superiority. It was essential that the roles were kept distinct. The fact that as one descended the ladder of white prosperity, one reached workers whose conditions of life were, in practice, little different from those of the slaves resulted

Cotton picking in America in the 1890s – a scene that could have been observed at any time in the nineteenth century.

in some of the more bitter forms of racial oppression. It was those at the very bottom of the ladder who usually felt the strongest need to impose a rung below that on which they themselves had landed. The fact that this superiority had no basis in reality had to be hidden behind an elaborate screen of pretence, half-truth and dubious morality.

The first justification offered by slave owners was that slavery had always been part of the African way of life. The Africans themselves bought and sold slaves. But even if it can be argued that an act can be justified simply because someone else does it – a catch-all morality that could be used to justify any crime – the argument would still have little force. Black slavery in America was of a very different kind from slavery in Africa. For the slave, it meant permanent exile to a foreign land, perpetual bondage not only for themselves but for their children as well for generations to come. There was not even the possibility of escape, for they carried with them the indelible mark of their slavery, the colour of their skin. To bring that state of affairs within the bounds of any morality required a neat twist of logic. The black skin that marked the slave for permanent bondage was the very thing that justified it. The colour itself proved inferiority. The argument was put crudely by crude men and more subtly by learned men, but its essence remained the same. Here, from among hundreds of different statements along the same line, is just one 'proof' of inferiority, in a paper read to the Anthropological Society of London in 1866:

The African has never reached … a higher rank than a king of Dahomney, or the inventor of the least fashionable *grisgris* to prevent the devil from stealing sugar plums. No philosopher among them has caught sight of the mysteries of nature; no poet has illustrated heaven or earth, or the life of man; no statesman has done anything to enlighten or brighten the links of human policy. In fact, if all that negroes of all generation have ever done were to be obliterated from recollection for ever, the world would lose no great truth, no profitable art, no exemplary form of life. The loss of all that is African would offer no memorable deduction from anything but the earth's black list of crimes.

From this argument only one conclusion could be reached:

50,000,000 of blacks have not been placed on this magnificent globe of ours for no purpose; it is therefore our duty, by wise legislation to utilize this large mass of human beings. They must be dealt with from no sentimental viewpoint, but from a knowledge of their nature and characteristics, discarding at once any theory of equality … One section must govern the other.

The argument of white supremacy could be buttressed on all sides. There was scientific mumbo-jumbo, so bizarre it is hardly possible to believe anyone ever took it seriously.

The negro's brain has in great measure run into nerves …from the diffusion of the brain, as it were, into the various organs of the body, in the shape of nerves to minister to the senses, every thing from the necessity of such a confirmation, partakes of sensuality, at the expense of intellectuality … The great development of the nervous system and the profuse distribution of nervous matter to the stomach, liver and genital organs, would make the Ethiopian race entirely unmanageable were it not that this excessive nervous development is associated with a deficiency of red blood in the pulmonary and arterial systems, from a defective atmospherization or arteriolization of the blood in the lungs.

This seems a rigmarole of nonsense, but the author, Dr Cartwright of New Orleans, took it seriously and his views were widely quoted as sound scientific fact. If you were not impressed with science, however, you could always turn to the most unimpeachable source of all, the Bible, which turned out to be full of texts to support the view that God was on the side of the white race. The black races were the children of Ham and had not Noah cursed him and his family: 'a servant of servants shall be unto his brethren.' No one seems to have queried just how presumably white Noah and his wife came to have a black son. If black was bad there was ample evidence that white was good. John's vision of the Second Coming could scarcely be plainer. 'His head and His hairs were white like wool, as white s snow.' Blacks then were intrinsically inferior and the white man was doing them a huge favour by bringing them out to enjoy the fruits of civilization.

What was more, he was offering them the Christian religion, which was so powerful it could turn the blackest skin white: but not, of course, in this life:

Here lies the best of slaves,
Now turning into dust.
Cesar, the Ethiopian, craves
A place among the just.
His faithful soul is fled
To realms of heavenly light,
And by the blood that Jesus shed,
Is changed from black to white.

This elaborate justification was applied to slavery because slave traders and slave owners found the system profitable. It was also highly speculative and the traders looked to get profits where and however they could. That might include a little privateering on the side, but the bulk of their wealth came from the triangular voyage: trade goods to the slave coast; slaves to America; and the produce of America back to Britain. The slave ship *Hawke*, sailing out of Liverpool in January 1779, took £3,000 worth of trade goods, mostly beads, brass, ironmongery and cloth. With these the slavers bought an unspecified number of slaves, of whom 386 survived the rigours of the crossing and were sold for over £17,000. At the end of the voyage, after paying wages and other expenses, the owners declared a profit of £7,000. The following year, she was even more successful, clearing £11,000, which included £3,700 for privateering, taking the French ship, *Jeune Emelia*, as a prize on the way home. In 1781, however, the tables were turned. The *Hawke* was captured and over £6,000 lost. Even so, the profit over the three years was still over £12,000. It would have been even higher but for the final misfortune. It was a trade that seemed worth any attempt to justify it being allowed to continue.

Needless to say, many never made any attempt at self-justification, and fewer still took the trouble to spell out their views at length. Nevertheless, it was a necessary process. It enabled men to live comfortably within a climate of general approval, and when the system came under attack from would-be reformers, it created a sense of a community acting on their own moral standards. The pro-slavery rhetoric had to sit uncomfortably by the realities of the trade. The high moral tone had little in common with the tone of the auction block, as this nasty little advertising jingle in the *South Carolina State Gazette* of September 1784 demonstrates:

Abraham Seixes
All so gracious,
Once again does offer

A slave auction in Virginia.

His service pure
For to secure
Money in the coffer.

He has for sale,
Some negroes, male,
Will suit full-well grooms.
He has likewise
Some of their wives
Can make clean dirty rooms.

For planting too,
He has a few
To sell all for the cash,
Of various price,
To work the rice,
Or bring them to the lash.

The slaves themselves, herded around the country like cattle, seemed inexplicably not to appreciate the favour that was being done to them by introducing them to 'civilisation'. J.W. Featherstonhaugh while journeying through the South wrote in *Excursions Through the Slave States* this description of a gang he chanced upon in 1844:

> It was a camp of negro slave-drivers, just packing up to start; they had about three hundred slaves with them, who had bivouacked the previous night *in chains* in the woods … It resembled one of those coffles of slaves spoken of by Mungo Park, except that they had a caravan of nine waggons and single-horse carriages, for the purpose of conducting the white people, and any of the blacks that should fall lame, to which they were now putting the horses to pursue their march. The female slaves were, some of them, sitting on logs of wood, whilst others were standing, and a great many little black children were warming themselves at the fires of the bivouac. In front of them all, and prepared for the march, stood in double files, about two hundred male slaves, *manacled and chained to each other.* I have never seen so revolting a sight before.

He noted that the gangs were constantly on the look-out for ways to escape, and he reported hearing of a recent incident where a slave had managed to get his hands on an axe, killing many of the overseers while the rest fled. Such accounts make nonsense of the often-repeated view that the black slave was essentially a submissive, docile creature, happy to accept his allotted place in life.

These traders were only links in the chain of men, passing on the slaves, and each making a profit – those who brought the slaves from Africa, those who sold them at auction, and those who carried them around the countryside. The firm of Templeman and Goodwin, for example, had their headquarters in Richmond, Virginia, and from there they made frequent expeditions into the other Southern states. Their book keeping was meticulous:

> 1849 Cost of Negroes taken
> Nov, 15th Martha cost 580 Sold to Absalon Dukes 625
> Isaac 675 Sold to Andrew Berry 825
> Caroline 655 Sold to Robt. F. Henderson 700

And so on. Within that month they sold nineteen slaves for a total of $12,950 and a $2,395 profit.

At the end of this trading, the slave had become an expensive commodity, and a valuable part of his owner's property. Now, like any other capital investment, he had to be turned into a profit. But however much slave owners might wish to regard the slave as a mere chattel, in the reality of everyday life he was no such thing. In theory, the owner could do with him just as he pleased but in practice, as we shall see later, the slave could impose limitations. There were other, intrinsic, limitations to the freedom of the owner wishing to turn his merchandise into a money maker. Work needed to be found, and agriculture provided the bulk of that work. There were to be later attempts to turn slaves towards manufacturing industry, just as the poor in Britain had been drawn into the factories. But the slave owner had problems that the mill owner never had to face. Responsibility for the slave was totally his, and the slave's well-being was essential for the protection of his investment. Mill hands could work for low wages and be left to look after themselves as best they might. There was no cash tied up in their persons, only in the machines at which they worked. The slave was different. He had to be kept healthy, fed, clothed, housed, kept in check and kept at work. The factory would not answer; the plantation would. Here the slave could grow the crops to feed themselves and their masters, and after that all that was needed was a suitable cash crop to provide the profit that was essential to the whole enterprise. Unless the owner could find that, the whole edifice would begin to crumble away. The slave was certainly a status symbol, but the status symbol had to earn his keep. Ambitious owners did not want a mere subsistence: they wanted cash in their pockets and the leisure time to enjoy it.

The main crops of the South were indigo, rice and, most importantly, tobacco. Virginia tobacco found a ready, but not infinitely extendable, market in Europe and prices fluctuated wildly. Planters grew so much tobacco that they were sometimes forced to burn their crops just to keep the prices high. It was a good cash crop, but the demand was just not high enough to support the entire plantation system. So the South looked around for an alternative, and their gaze turned to the hungry mills of industrial Britain.

The attractions of cotton were plain to see. It suited the climate, there was a ready and expanding market and it was ideal for the big fields of the plantations. The first cotton fields were planted along the eastern seaboard, but soon spread inland. The West Indies planters faced by this competition simply shifted their ground. They had a new crop, sugar cane, which promised to do well for them, so they left cotton growing to the Americans. There was, however, one obstacle to expansion. To be acceptable to British manufacturers, the cotton had to be supplied in good, clean condition and free of seeds. It was failure to meet these conditions that had led to the decline in the Indian trade. Cleaning seeds from cotton is not easy – they cling hard to the fibres. They could be pulled out by hand, but this was a tedious and time-consuming business. They were also removed by 'bowing' in which a strung bow was vibrated in the cotton, sending the seeds flying out. But neither method was very efficient, nor were they suited to the shorter staple cotton now being favoured in the South. The obvious answer was to invent a machine to do the job, for if cotton could be spun by machine it should not be beyond the ingenuity of man to find a way to mechanise the cleansing. So it proved, but the answer was not found by a native Southerner with experience in cotton growing, but by a visiting Yankee.

Eli Whitney, son of a Massachusetts farmer, arrived in the South, having graduated from Yale in the Autumn of 1792. He came to Savannah, Georgia, and from there travelled up-river to stay on the plantation of Mrs Greene, widow of General Nathaniel Greene, who had been awarded the property for his gallantry during the Revolutionary War. Young Eli, during the same war, had shown that he had more of a taste for practical work than academic when he set up a small forge to manufacture nails. Now in the South he could not help but hear the complaints about cotton cleaning, how it could take a whole day to get one slave to clean just a pound of fibres. He decided to solve the problem himself and set up a workshop for his experiments. As with so many inventions of this time – James Watt watching a kettle boil and going on to invent a steam engine is a classic example – it seems the world likes to wrap reality in a romantic story, rather than simply admit that one man worked and thought hard to come up with a solution. In Whitney's case, the legend is that he was watching a farm cat sitting outside the poultry yard, protected by a chicken-wire fence. When one of the chickens got close enough, a paw shot out – a partly bald hen scurried off and the cat was left with claws full of feathers. That is the story, and it does indeed explain the principle behind the invention. Cotton was picked up by a roller, studded with metal teeth that carried it round to a metal grille. The fibres were scraped off and the seeds dropped away. An improved version by Hogden Holmes replaced the

The Whitney gin: a simple device that revolutionised the cleaning of cotton.

The prosperous plantation followed on from the invention of the Whitney gin.

studded roller with a circular saw blade. Simple and effective, it was named the cotton gin and was the making of the South. Now one slave turning a handle could do the work of a dozen, and larger gins could be built to be worked by water wheels, so that they could do the work of a hundred. The solution had been found, and Whitney sat back to collect the rewards.

The story of the Whitney gin is very similar to that of other inventions in the British textile industry. He went into partnership with a man called Phineas Miller and attempted through patents to get a monopoly of gin manufacture. Their prices were high, but the machines were simple. Plantation owners much preferred building their own gins, without any payment to Whitney. So many followed this route that it was impossible to pursue them all for patent payments. Eventually, the government stepped in and bought out the patent for $50,000. Whitney went back North, where he devised a system for making muskets from standardised parts – the basis of mass production. It was a major technological breakthrough, but of no interest for the South. They had got what they wanted from Whitney – a way to make cotton growing seriously profitable.

Chapter 6

TOWARDS THE FACTORY AGE

Cotton supplies to Britain seemed assured, and the impetus given to production by the invention of the water frame and the establishment of mills on the Cromford pattern was given a powerful boost by yet another important invention. The yarn from Arkwright's spinning machines was comparatively coarse, and manufacturers needing finer yarn still had to rely on it being produced by the less efficient jenny. That situation changed when Samuel Crompton introduced his spinning mule. A weaver, born into a comparatively well-off family, Crompton hit on the idea of combining the principles of the jenny and the water frame in a single hybrid machine – hence the name 'mule'. It used rollers to draw out the yarn, which was then passed on to spindles, mounted on a moveable, wheeled carriage. As the carriage retreated, the yarn was stretched out further, and when the carriage stopped, the spindles were rotated to impart spin. A bar then dropped onto the threads and the carriage began the return journey, winding the cotton onto bobbins.

Crompton built his first machine at his home at Hall-i'-th'-Wood in 1779, just at the time that the spinners of Lancashire were rampaging through the region, burning mills and trashing machines. He decided, very reasonably, that this might not be the best time to introduce another new machine, so he dismantled the protype and hid it away. Once the trouble had subsided, he put the mule together again and set it to work. At this time, all he had in mind was establishing an advantage over his neighbours by supplying fine yarn at low cost, without mentioning how he had managed it. The difference in price, however, was so marked that it was obvious he had found a new spinning method and manufacturers were soon hammering on his door, wanting to be let into the secret. The familiar inventor's story was about to be re-enacted.

Crompton listened to his tempters, believed what they said and offered his invention to the world at large. The local manufacturers had promised handsome compensation but had somehow never got round to mentioning an actual figure. Eighty-five of them entered into

Samuel Crompton, inventor of the spinning mule.

the agreement with Crompton and when all their subscriptions were gathered in, £72 was the final miserable total. Crompton struggled on, watching others making fortunes from his invention, and at last petitioned parliament in the hope of getting some reward for his efforts. Many leading manufacturers gave evidence on his behalf to the Select Committee of the House of Commons, and he had a firm ally on the Committee, the successful manufacturer Sir Robert Peel. His luck seemed to have turned. He was talking to Peel in the lobby of the House, when they were approached by the prime minster, Spencer Perceval, with the good news that he was sure that Crompton was going to be awarded £20,000. The date was 1 May 1812, and that afternoon Perceval was assassinated. Parliament was dissolved and poor Crompton's hopes were dissolved with it. He died, as so many others had done, an embittered man, who had to sit helplessly watching others make the profits and leave him with nothing.

As the eighteenth century ran its course, so the pace of development quickened – and continued to accelerate. The manufacturers now had it within their power to supply any quality of yarn by means of the new machines established in factories. Once the turmoil of 1779 died down in Lancashire, the mill builders returned with new enthusiasm, until it seemed that any stream that could turn a waterwheel was lined with new buildings. And as the pressure on water supplies grew, so the manufacturers spread out to other regions – to

Spinning mules at work: note the small girl under the machine, cleaning up.

Scotland and the English Midlands, where the burgeoning hosiery trade provided a ready market. With each new mill that was built, more and more were pulled into the factory system and yet more bustled round them to get their share of the profitable trade. Weavers established their workshops round the spinning mils; transport systems grew to bring in the raw materials and take out the finished products; canals were cut and roads improved. Hamlets grew into villages and villages into towns with shops, inns and traders to serve the new communities. Travellers from the south to northern England came back with stories of changes at a pace they would never have thought possible. Some saw the process as part of a triumphant march of progress, heading to a Utopian future. As Arthur Young puts it in *A Six Month Tour Through the North of England* in 1769.

> 'Get rid of that dronish, sleepy and stupid indifference, that lazy negligence, which enchains men in the exact path of their forefathers without enquiry, without thought and without ambition, and you are sure of doing good.'

Others looked on woefully at the destruction of the old order, in which they had held a place made comfortable by centuries of tradition that was threatened by the new: 'Every rural sound is sunk in the clamour of cotton works; and the simple peasant is changed into the impudent artisan.'

Whether they applauded or condemned the changes, all agreed they were momentous. The commentators, however, all had one thing in common: they were wealthy and privileged. They tended either to be aristocrats concerned with maintaining the status quo or industrialists beginning to demand a voice in the nation's affairs. The latter applauded the end of the old stagnation and looked forward to a dynamic economy creating fresh wealth and opportunities. The wealth was not, however, to be made available to everyone. Aristocrat or prosperous middle class, both agreed in their approach to the lower orders. The aristocracy, shaken by revolution in France, distrusted and even feared the lower classes. They had to be kept in their place – and that place meant they had to be kept poor. The new class who were paying the wages had no problem with that attitude. Being poor meant they had to be industrious to survive. In their ideal world, the poor would be kept far too busy working in the mills to have time to think about revolution.

Viewed in such a light, the status quo was maintained. The poor did remain poor, but not quite as poor as they had been before industrialisation swept across their world. Looked at from another point of view, the changes were not just about material wealth. Man may not live by bread alone, but it does help if he has the bread and enough money left over to put jam on it. But the changes brought about by the growth of the cotton industry affected every aspect of the workers' lives.

To understand these changes, one has to look at what the cotton operatives might have been doing had the mills not existed – and what life was like in other parts of the country still untouched by the changes. A good starting point is the sector that was still the largest employer of labour in Britain – agriculture.

Throughout the period of violent change in industry, similar upheavals were changing the world of the farmer and the farm labourer. The land was being improved and producing better yields. Cattle were getting fatter and healthier – the average weight of cattle sold at Smithfield Market in London in the 1790s, for example, was twice that of the beasts sold in 1710. But improvements were bought at a price, and much of it was paid in the enclosure movement. What had once been common land where villagers could graze their own livestock, was taken into private ownership. Fences were erected and hedges planted, behind which the new landowner could set about bringing in new and better crops. But compensation was seldom paid to the people who had lost their old rights and they resented it as expressed in a rhyme of the time.

The fault is great in man or woman
That steals the goose from off the common.
But what can plead that man's excuse
That steals the common from the goose.

There was now, however, surplus food to supply the new towns, and profits to be made from the surplus. The enclosure movement expanded at an astonishing rate. In the first half of the eighteenth century, 115 Enclosure Acts were passed by parliament; in the second half, that number rose to 2,015. You could claim compensation if you could prove you had some rights to the land being enclosed but the vast majority simply relied on unwritten tradition. The rights to common grazing and gleaning may have played a comparatively small part in the lives of those who made their living as labourers, but in a subsistence economy that difference was crucial. The dispossessed had to fight for their rights as best they could, but the law was always on the side of properly laid out documents, not vague tradition. There were protests. The men of Cheshunt in Hertfordshire, for example, made their views plain in a letter to the Squire of Cheshunt Manor, Oliver Cromwell, in 1799:

Whe write these lines to you who are the Combin'd of the Parish of Cheshunt in the defence of our Parrish rights which you unlawfully are about to disinherit us of … Resolutions is maid by the aforesaid Combin'd that if you intend of inclosing our Common fields Lammas Meads Marshes & Whe Resolve before … that bloudy and unlawful act is finished we have your hearts bloud.

They never did resort to violence and managed to obtain a miniscule sum as compensation.

The rural poor were seldom depicted realistically by painters, except as here where they were intended to show the results of indolence. The urban poor often lived in even worse conditions.

The independent yeoman of folklore became the unlanded labourer, working from dawn to dusk for a pittance that all too often fell short of meeting even the most basic family demands. Paid less than a living wage, the labourer and his family were too often forced into the humiliation of begging relief from the parish. Eden's report on the condition of the poor gave wages to the farm labourers as low as 1s a day. And what would that have bought the worker after he had set aside money for rent? Eden noted that meat was 4½d to 5d a pound, wheat 12s a bushel, butter 13d a pound and cheese 4d. The shilling, it seemed, would buy very little and the magistrates at Speenhamland laid down a rule that a subsistence level should be agreed and that where wages fell below that, the parish would make up the difference. They meant well, no doubt, but it was a direct incentive to employers to pay low wages. It was a harsh rule that stripped hard working men of their dignity. It was a bad rule, yet sensible men saw it in operation, noted its results but failed to condemn it. They offered instead the cheerless doctrine of hierarchies which said that some were born to rule, others to serve and it was blasphemy to try and interfere with this divine order, as Sir Frederick Eden explained:

> The Poor have nothing to stir them up to labour but their wants, which it is wisdom to relieve, but folly to cure. The maxim is not less calculated for the advantage of the Poor, than it appears for the benefit of the Rich. For, among the labouring people, those will ever be the least wretched to themselves, as well as most useful to the Public, that being meanly born and bred, submit to the station they are in with cheerfulness; and contented that their children should succeed them in the same low condition, inure them from their infancy to labour and submission, as well as the cheapest diet and apparel.

The author of these fine sentiments then proceeded to lecture the housewife on just what might constitute that 'cheapest diet':

> A pound of good beef or mutton, 6 quarts of water, and 3 ounces of barley, are boiled till the liquor is reduced to about three quarts: one ounce of oatmeal, which has been mixed up with a little cold water, and a handful or more of herbs, are added.

This he claimed should last a large family for three days. These conditions were not even the worst that could be found. In Scotland, for example, in the 1790s, a cash wage of 1s a day was the top rate. In the impoverished Highlands, things

Calico printing machines in Britain imitated the hand-painted cottons of India.

were even worse than in the Lowlands. There the labourer was given food at work, but his wages were reduced to a mere 8d. In all regions, however, wages were lower in winter than in summer.

There is no need to go on making the same point. Life in the fields was brutal and hard, with starvation an ever present threat. At best the labourer might earn his keep; at worst he was forced to go cap in hand to the parish. And if that occurred, he was liable to be faced with the compulsory break-up of his family, with children being sent away as apprentices to the new mills. This was the climate of opinion in which the conditions in the mills was set and the gloomy background against which they can be judged.

Chapter 7

MILL CHILDREN

The children who came from all over the country to serve in the new mills were mainly parish apprentices. Some were orphans, others children of labourers whose families could no longer afford to keep them. Because the mills controlled all aspects of their lives, the rules they enforced and the conditions they provided give a good insight into employers' attitudes. The records of one mill in particular give a very complete account of the way of life of the children. It was generally regarded as one of the best and most humanely organised in the whole country.

Samuel Greg of Belfast came to the hamlet of Styal in Cheshire, just a few miles south of Manchester, in 1784. There he built a large mill on the banks of the Bollin, and a short distance away he built a village to house the workers he would recruit for the new factory. The adult workers signed up for varying periods of time. James Stretch of Morley came along for three years 'at the wages that Mr. Greg gives to other spinners'; and if he left before that time, he agreed to pay Greg one guinea 'for learning to spin'. Others had a more dubious background. Daniel Bate, a clockmaker from Middlewich, was no doubt glad enough to get out of the Manchester House of Correction to work for five years at 15s a week – though first he had to earn and hand over the guinea that Greg had paid to get him released. Whole families arrived and signed up for work, often borrowing from Greg to help them get started in their new homes. Most of the families came from Cheshire or the neighbouring counties of Lancashire and Staffordshire. Some of the children who came as parish apprentices were local, but many came from much further away. Two popular sources were the Poor Houses of Chelsea and Liverpool. Parishes were only too eager to unload their parish children.

> The thought has occurred to me that some of the younger branches of the poor of this parish might be useful to you as Apprentices in your factory at Quarry Bank. If you want any of the above, we could readily furnish you with Ten or more at from nine to twelve years of age of both sexes.

Greg replied, setting out his terms, for relieving the parish of these burdens:

> I am much obliged by your attention and find we have room at present for about 12 young Girls from 10 to 12 years old … the terms at which we take them are:

> Two guineas each will be expected from the Parish, and clothing sufficient to keep the Children clean.

Say 2 shifts 2 frocks
2 Brats or aprons
And two guineas to provide them with other necessities

The parish must have thought it was doing well to get rid of the expense of keeping the children for a few guineas, while Greg was equally pleased with the bargain. Indentures were drawn up setting out the liabilities on both sides:

It is this day agreed by and between Samuel Greg of Styal in the County of Cheshire, of the one part and Sarah Irwin Daughter of John Irwin of Newcastle on the on the other Part, who Agrees to the terms as follows. That the said Sarah Irwin shall serve the said Samuel Greg in his cotton mill at Styal, in the County of Cheshire, as a just and honest Servant, Twelve hours in each of the six working days, and to be at her own liberty at all other times; the commencement of the Hours to be fixed from Time to Time by the said Samuel Greg, for the Term of four years at the Wages of one Penny p week allso sufficient meat drink apparel washing and other things necessary and fit for one in her situation.

Further clauses specified that if she was absent from work Greg 'may abate the Wages in a double Proportion' and that Greg could get rid of her 'for Misbehaviour or Want of Employ'. Agreement reached, the young girl was brought to Styal and given a new home in the apprentice house.

'Here are well-fed, clothed, educated, and lodged, under kind superintendence, sixty young girls, who by their deportment at the mill … evince a degree of comfort most credible to the humane and intelligent proprietors.'

The apprentice house still stands and would not be thought particularly grand as an ordinary family home.

The description was by the Pangloss of the factory system, Andrew Ure, in his book *The Philosophy of Manufacture*, 1935. If he ever saw anything amiss in any mill in the land that he visited, he never mentioned it. The children themselves, far from home, working among strangers for twelve hours a day, wrote no books to express their views. They could – and did – make their feelings known more directly. Joseph Stockton, a penny-a-week apprentice from Newcastle-under-Lyme, was signed up for eight years in 1796. On 12 May 1799, he ran away and was tracked down to Newcastle. A warrant was issued and he was hauled back to Styal on 10 June. Next day he was off again and was caught by the constable on the Newcastle road. On 13 June 'he promised to mind his business and give us no further trouble in future and was again set to work'. That lasted just over a month and on 22 July he ran off for the third time. There is no further mention of him in the records. Why did he run away? The records do not say. What happened to him? Again, no answer. Perhaps he finally escaped, or found himself in gaol. Two other runaways, however, give us rather more information about life in the mill. Joseph Sefton and Thomas Priestley made it all the way to London, in Joseph's case because he wanted to see his mother. They were caught and brought in front of the magistrates in August 1806. In their evidence they gave a very complete picture of their life at Styal.

Joseph Sefton. I am 17 years of age this Autumn. My father I am informed deserted me or went for a soldier when I was about 2 years old. His name was John Sefton. I was told that I was born at Clerkenwell I had been in the workhouse of the Parish of Hackney from an infant about 3½ years I consented before the magistrates to be bound apprentice to Samuel Greg cotton spinner and manufacturer. There were 8 boys and 4 girls of us bound at the same time. We went to Styal and were employed in the cotton mill of Mr Saml Greg which was a short distance. I was first employed to doff bobbins [remove bobbins full of yarn and replace them with empty ones] … I used to oil the machinery every morning. In fact I was employed in the mill work. I did not spin. I liked the employment very well. I was obliged to make over time every night but I did not like this as I wanted to learn my book. We had a school every night but we used to attend once a week (besides Sundays when we all attended) … I wanted to go oftener to school but Richard Bamford would not let me go … I have no reason to complain of my master Mr Greg nor Richard Bamford who overlooks the work there were 42 boys and more girls apprenticed. We lodged in the Prentice House near the Mill. We were under the care of Richard Sims and his wife. The boys slept on one side of the house and the girls on the other. The girls all slept in one room, the boys in three. There was a door betwixt their apartments that was locked of a night. Our rooms were very clean, the floors frequently washed, the rooms aired every day, whitewashed once a year. Our beds were good. We slept two in a bed and had clean sheets once a month. We had clean shirts every Sunday. We had new clothes for Sunday once every two years. We had working jackets new when those were worn out and when our working trousers were dirty we had them washed. Some had not new jackets last Summer but they were making new ours when I came away.

On Sunday we went to church in the morning and to school in the afternoon after which we had time to play.

On Saturdays we had for diner boiled Pork and potatoes. We had also peas turnips and cabbages in their season.

Monday we had for dinner milk and bread and sometimes thick porridge. We had always as much as we could eat

Tuesday we had milk and potatoes

Wednesday sometimes Bacon and Potatoes sometimes milk and bread

Thursday if we had Bacon on Wednesday we had milk and bread

Friday we used to have Lobs scouse

Saturday we used to dine on thick porridge

We had only water to drink. When ill we were allowed tea.

Thomas Priestley was 13 years old, and two months before he ran off, he suffered one of those accidents that were all too common among the unguarded machinery. 'I was working and there was a great deal of cotton in the machine, one of the wheels caught my finger and tore it off … I was attended by the Surgeon of the factory Mr Holland and in about six weeks I recovered.' His deposition gives more details of the working day.

> Our working hours were from six o'clock morning Summer and Winter till 7 in the evening. There were no nights worked. We had only ten minutes allowed us for our breakfasts which were always brought to the Mill to us and we worked that up at night again – 2 days in the week we had an hour allowed us for dinner, while the machines were oiled, for doing this I was paid a ½d a time, on other days we were allowed half an hour for dinner. When the boys worked over time, they were paid 1d an hour.

Cruikshank's view of brutality in the cotton industry. The large bale being carried out labelled Sir R. Peel is a reference to the fact that Sir Robert Peel, who became prime minister, had made a fortune in the industry.

In August 1806, there were ninety apprentices in the modestly sized apprentice house, and numbers frequently rose above that number. Life at Styal was considered easy by the standards of the time, and other apprentices paint an even bleaker picture of their time at other mills. *A Memoir of Robert Blincoe* was published in 1835 and in it he described how he was taken from the St Pancras workhouse in London at the age of seven as an apprentice in a Derbyshire mill. In 1815, at the age of 10, he was one of 100 apprentices, boys and girls, in the apprentice house at Litton Mill, established by Ellis Needham in 1780. There was no resident supervisor and the house was like a prison, surrounded by a high stone wall, kept locked at night.

We all ate in the same room, and all went up a common stair to our bed chamber, all the boys slept in one bed chamber, and all the girls in another. The beds were in rows along the wall, a second tier being fixed above the first. The beds were thus made double by the square frame work – one bed above, the other below. This was done to save room. There were about twenty of these beds, and we slept three in one bed. The girls' bed-room was of the same sort as ours. There were no fastenings to the two rooms and no one to watch over us in the night or to see what we did.

We went to the mill at five o'clock without our breakfast, and worked till about eight or nine, when they brought us our breakfast, which consisted of water porridge with oatcake in it and onions to savour it with, in a tin can. This we ate as best we could, the wheel never stopping. We worked on till dinner time, which was not regular, sometimes half-past twelve, sometimes one. Our dinner was thus served to us. Across the doorway of the room was a cross-bar like a police bar, and on the inside of the bar stood an old man with a stick to guard the provisions. These consisted of Derbyshire oat cakes cut into four pieces, and ranged in two stacks. The one was buttered, the other treacled. By the side of the oat-cake were cans of milk piled up – butter-milk and sweet-milk. As we came up to the bar one by one the old man called out 'Which'll 'ta have, butter or treacle, sweet or sour?' We then made our choice, drank down the milk and ran back to the mill with the oat-cake in Hand, without ever sitting down. We then worked on till nine or ten at night without bite or sup. When the mill stopped for good, we went to the house for our supper, which was the same as breakfast – onion porridge and dry oat-cake.

It makes Styal seem almost idyllic by comparison and it was not the worst. There were mills where work continued day and night. As in so many ways, Arkwright's Cromford Mill led the way, introducing gas light at a very early date. As one group of apprentices fumbled their way out of bed, another set walked wearily in to take their place. It was then that an old saying was born: 'The beds of Lancashire never got cold'.

Some relief was offered by the Peel Act of 1802 that gave protection to the pauper apprentices. It limited working hours to twelve a day and stipulated that

Litton Mill in Derbyshire, where Robert Blincoe served his apprenticeship.

no apprentices should do nightwork. It was an acknowledgement that something had gone badly wrong and that someone was prepared to do something about it, but not very much. In any case, the paupers only represented a portion of the whole apprentice population and the passing of the Act meant very little as, although two inspectors were required to be recruited for each district, they had no authority to impose penalties, so in effect they had no power to enforce the law. John Moss, an overseer at the Beckbourn Mill apprentice house near Preston, gave evidence before the Committee set up in 1810 to enquire into the working conditions of the factory children. His evidence told of children working from 5.00 am in the morning to 8.00 pm, or even longer if there was time to make up. One hour out of the day was set aside for meals. Their only rest came on Sundays, when they only had to work from 6.00 am to noon, cleaning the machinery. As the Committee put their questions a sad story unfolded:

Did the children sit or stand at work? – Stand

The whole of their time? – Yes.

Were there any seats in the mill? – None.

Were they usually much fatigued at night? – Yes, some of them were very much fatigued.

Did you inspect their beds? – Yes, every night.

For what purpose? – Because there were always some of them missing, some sometimes might have run away, others I have sometimes found asleep in the mill.

When asked why he had so blatantly disregarded the conditions set down in the 1802 Act, he replied quite simply that he had never heard of it, and there was no reason to suppose he had lied. Many another overseer must have gone on in the old ways simply because he was ignorant of the law or certain that no one would try to enforce it.

By 1816, the Act was already not only ineffective but largely irrelevant. Manufacturers were moving away from the apprentice system to the more straightforward alternative of hiring children as ordinary wage-earners. It could be argued that the change was the direct result of the Peel Act, an attempt to get round the restrictions on working hours. It is, however, much more likely that mill owners were discovering that the apprentice system was uneconomical. Samuel Greg calculated in 1790 that the average weekly cost of keeping an apprentice was 3s 6d and by 1822 this had risen to 5s. Wages were a lot cheaper than that, and with the change the owners were relieved of any responsibility for the children's

A young girl mixing colours for a block printer.

welfare. The decline of the apprentice system, however, did not mean any great change in working conditions.

Within this picture of gloom, there were patches of brightness and they shone brightest in Scotland, where a novel doctrine was being put forward. Robert Owen of New Lanark, one of the most important manufacturers in the country, addressed these words to his fellow industrialists:

> Will you then continue to expend large sums of money to procure the best devised mechanism of wood, brass, or iron: to retain its perfection; to provide the best substance for the preservation of unnecessary friction, and to save it from falling into premature decay? Will you also devote years of intense application to understand the connexion of these various parts of these lifeless machines, to improve their effective powers, and to calculate with mathematical precision all their minute and combined movements? And when in these transactions you estimate time by minutes, and the money expended for the chance of increased gain by fractions, will you not afford some of your attention at consider whether a portion of your time and capital would not be more advantageously applied to improve your living machines?
>
> From experience which cannot deceive me, I venture to assure you, that your time and money so applied, would return you not five, ten, or fifteen per cent of your capital expended, but often fifty and in many cases a hundred per cent.

Owen put forward the revolutionary view that the humans in the mill mattered in his book *A New View of Society*, which was based on his experience at the New Lanark mill. The mill had been established by Arkwright and David Dale, and Owen arrived in 1798, fifteen years after its foundation. As at Styal, the workforce was divided between the families housed in new tenement blocks, let out by Dale at very low rents, and some 500 pauper apprentices, mostly from Edinburgh. He found the village and villagers fell some way short of the ideal. 'The population lived in idleness, in poverty, in almost every kind of crime; consequently in debt, out of health, and in misery.' At first sight, the condition of the children was far better:

> The benevolent proprietor spared no expense to give comfort to the poor children. The rooms provided for them were spacious, always clean, and well ventilated; the food was abundant, and of the best quality; the clothes were neat and useful; a surgeon was kept in constant pay to direct how to prevent or to cure disease; and the best instructors which the country afforded were appointed to teach such branches of education as were deemed likely to be useful to children in their situation. Kind and well disposed persons were appointed to superintend all the proceedings.

The streets of New Lanark, c.1900. The workers' tenements line the street that leads up to the court house.

Appearances were deceptive. The authorities insisted that Dale took children as young as 6 years of age, and Dale found it necessary to work these little children from 6.00 am to 7.00 pm. He might offer them educational opportunities, but what use was that after thirteen hours crippling labour in the mill – 'many of them became dwarfs in body and mind, and some of them were deformed' – it is an appalling indictment of the system that that was the end result of the best efforts of one of the kindliest employers.

Owen set about a vigorous improvement campaign. He took on the role of benign despot, cajoling, instructing and demanding cleanliness, sobriety and honesty. Housewives would find Owen on their doorstep insisting on inspecting their handiwork. The worst of the ale houses were closed and Owen endlessly preached the virtues of temperance. He put the greatest emphasis on education. No building was so important, in his eyes, as the school he set up – the New Institution. Here both adults and children were taught. But how did he succeed where Dale had failed? Owen had a revolutionary answer. Children under 10 were to attend school and would not be allowed to take on mill work. If the local authorities refused to send pauper children, then so be it; there would be no pauper apprentices at New Lanark. But could adults and the older children be expected to benefit from education when they emerged exhausted from the mill?

Clearly they could not, so hours would have to be reduced to the point where they were not exhausted.

These were wildly controversial views and they did not go unopposed. Owen's partners in the enterprise viewed with alarm the prospect of being dragged willy-nilly down the path to bankruptcy by the passionate reformer. But Owen was not to be dissuaded. What was better for the long-term good of the concern, he demanded to know, a healthy, hard-working employee or a wretched hand, demoralised and dishonest? The experiment was tried and it worked. Owen was able to demonstrate that shorter hours produced more goods and of a higher quality, and he had the figures to prove it. A healthy individual did produce more than one stumbling to the point of exhaustion.

Such views were greeted with disbelief and ridicule when Owen was called to give evidence to the Peel Committee of 1816. The members kept asking him over and over again to explain the rise in productivity. Surely, they insisted, he must be speeding up the machines. No, replied Owen, healthy, alert operative make fewer mistakes so there are fewer stoppages, fewer delays. The Committee simply could not accept the explanation. 'Do you', they demanded, 'as an experienced spinner, or spinner of any kind, mean to inform the Committee that the machine that you employ for throstle and water spinning can produce an additional quantity from any other cause whatever but the quickening of the motion of the machine?' Owen could only repeat his answer, and the Committee went on repeating the question until at last Owen was forced to reply, rather tetchily, that 'it is far from my wish to deceive the Committee'. At the end of the day, they were no more convinced than they had been at the beginning and other manufacturers shared their scepticism.

The New Lanark experiment remained an isolated affair, a bright lead that no one wanted to follow. Not, of course, that everything in New Lanark was beyond reproach. Owen wanted to see an industrial society that reflected what he saw as the finest elements of previous ages. He wanted harmony between masters and employees, all working together for the common good. Co-operation was the key word, but co-operation along lines that he laid down. He was an autocrat, demanding obedience from the workforce not just in the mill but in their private lives as well. Not every rule he laid down was ungrudgingly accepted, nor obviously to the advantage of the workers. He was a fanatical believer in the virtues of music and dance and tenants were dragooned into taking part. One woman eventually packed up and left, complaining that Owen 'had got a number of dancing masters, a fiddler, a band of music, that there were drills and exercises and that they were dancing together till they were more fatigued than if they were working'. But whatever his quirks, it was by his treatment of the children that Owen should be judged. Among the literally thousands of visitors who came to see this fascinating experiment was the poet, Robert Southey, who watched the scene in the infants' playground with pleasure:

> It was really delightful to see how the little creatures crowded about Owen to make their bows and their curtsies, looking up and smiling at his face, and the genuine benignity and pleasure with which he noted them ... The shouts and laughter of the children were worth all the concerts of New Lanark, and of London to boot.'

In this, at least, Owen's despotism seems justified.

In reading the seemingly endless accounts of overwork, brutality and near enslavement of the mill children, one question inevitably arises – why did almost no one take up their cause? In part one can find an explanation in the ethos of the age that held the view that the greatest gift you could bestow on the poor of any age was to provide them with work. W. Hutchinson, visiting a carpet factory at Penrith in 1776 recorded in *An Excursion to the Lakes* that 'tears of pleasure gushed upon the eye' when he saw the apprentices at work. By that work they were 'saved from the hands of destruction and vice, rendered useful members of society, and happy in their industry and innocence.' The manufacturer, thinking perhaps more of a good return on his capital, was, nevertheless, delighted to receive justification from the old order of landed gentry. The landowners were, in any case, not well positioned to attack the working conditions of factory children. The child of a farm labourer soon found it had to earn its crust. As soon as they

The children of New Lanark putting on a dancing display for visitors to the New Institution, c. 1825.

were old enough, they were set down at a gate at the end of the lane, according to William Howitt in *The Rural Life of England* (1838):

> ...there through long solitary days they pick up a few halfpences by opening it for travellers. They are sent to scare birds from corn just sown, or just ripening ...They help to glean, to gather potatoes, to pop beans into holes in dibbling time, to pick hops, to gather up apples for the cider mill, to gather mushrooms and blackberries for market, to herd flocks of geese or young turkeys, or lambs at weaning time ... to bring in wood for the fire, or to rear turfs for drying on the moors, as the man cuts them with his paring shovel, or to rear peat-bricks for drying. They are mighty useful animals in their day and generation, and as they get bigger, they successively learn to drive plough, and then to hold it; to drive the team, and finally to do all the labours of a man.

Were factory children then any worse off than farm children? It's tempting to answer with an immediate 'yes'. At least the farm children were out in the open air, not stuck away in a dark, unhealthy mill. But the image of open-air life that probably comes to mind is one of fields of grain in the sunshine, not of February mornings when frost turns the ground to iron or November mists when the clinging damp chills to the bone. But whatever the hardships of rural life, it still seems preferable to that of the factory.

Children were beaten in factories, and, so no doubt, were some on farms. Factory children worked long hours, but so did those on the land. Both were touched by disease and the miseries of poverty. Yet crucial differences remain. The factory child was taken from a familiar environment, clamped within an iron discipline and a regime untouched by the changes in the seasons. Winter was as summer, the working day was measured by the tick of the clock, not the movement of the sun. Each day was as the last – no time of sowing, no harvest home. And in every minute of every working day, the machines ground on, determining the rate of work, giving no room for personal decisions and choices. There are many harrowing accounts in the evidence presented to the 1833 Royal Commission on the work of children in factories. Often removed from all family ties, the factory child was at the mercy of the overseer who, too often, was forced to keep the children at work by whatever means he could devise:

> After the children from eight to ten years had worked eight or nine or ten hours, they were nearly ready to faint; some were asleep; some were only kept to work by being spoken to, or by a little chastisement, to make them jump up. I was sometime obliged to chastise them when they were almost fainting, and it hurt my feelings; then they would spring up and work pretty well for another

Children employed at Coat's of Paisley in the late nineteenth century.

hour; but the last two or three hours were my hardest work, for then they got so exhausted.

It was an experience that was almost as bitter and dehumanising for the conscientious overseer as it was for the children. Parliament was eventually to legislate against the worst excesses, but by then the factory system had already been in operation for more than sixty years. For generations of children, the legislation came too late. And all the time, the industry was spreading, more and more mills were being built. To the economist it was a sign of economic growth: to others it meant misery. But to the very poorest in Great Britain, and especially in Ireland, it did seem to offer a way out of extreme poverty. They crowded into the textile districts. Mills became bigger, towns grew larger and the slums became deadlier.

Chapter 8

COTTON TOWNS

The cotton manufacturers armed with water frame, jenny and mule might seem to have all they needed for an unopposed advance towards a previously unimaginable prosperity. There were, however, a few difficulties still to overcome. There were investors in plenty ready to put their money into new mills, but they needed the right site to build them on. Ideally, they would be next to one of the new improved turnpike roads or better still beside a canal or river navigation, but there was one other criterion that had to be met. They needed a reliable water supply to turn their wheels, and such prime sites were disappearing fast. Mills were being built some distance from the best transport routes and centres of population. Ideally, what they wanted was a new source of power. They were about to get one.

The first steam engines had come into use, in one form or another, as early as the beginning of the eighteenth century, but they were only designed to pump water, mostly from mines. It was James Watt who made the great breakthrough by designing a steam engine that could not only be used to lift pump rods up and down but could be adapted to turn a shaft. It could, in other words, take the place of the water wheel. And, in 1778, the first Boulton and Watt steam engine was installed in Robinson's mill at Papplewick, near Nottingham. This was good news not just for prospective mill builders but also for existing manufacturers who, in the past, had found work halted for lack of water during a drought. At Styal, for example, they installed an engine quite early on. Here, as in many other mills, it was a belt-and-braces operation. If one source of power failed, another was ready to take over. It was comforting to have the steam engine during a dry spell – and it was equally comforting to have the old, reliable water wheel if the new-fangled machine broke down. If the introduction of the steam engine had meant no more than this, then it would have been just another useful addition to the mill's battery of machines, but it was to have far greater significance.

The new steam-powered mills could be put almost anywhere, and manufacturers knew just where they wanted them, in the middle of a town and close to a major transport route, with the workforce literally on the doorstep. No need now for horses to plod down a muddy track to reach a distant mill; no need either for workers to waste energy – and time – in tramping between home and work. Everything could be gathered together. So, the cotton towns grew up around the new mills – Bolton, Blackburn, Oldham and the rest. By 1811, there were forty-two spinning mills in Oldham alone, and around them the new houses were clustered

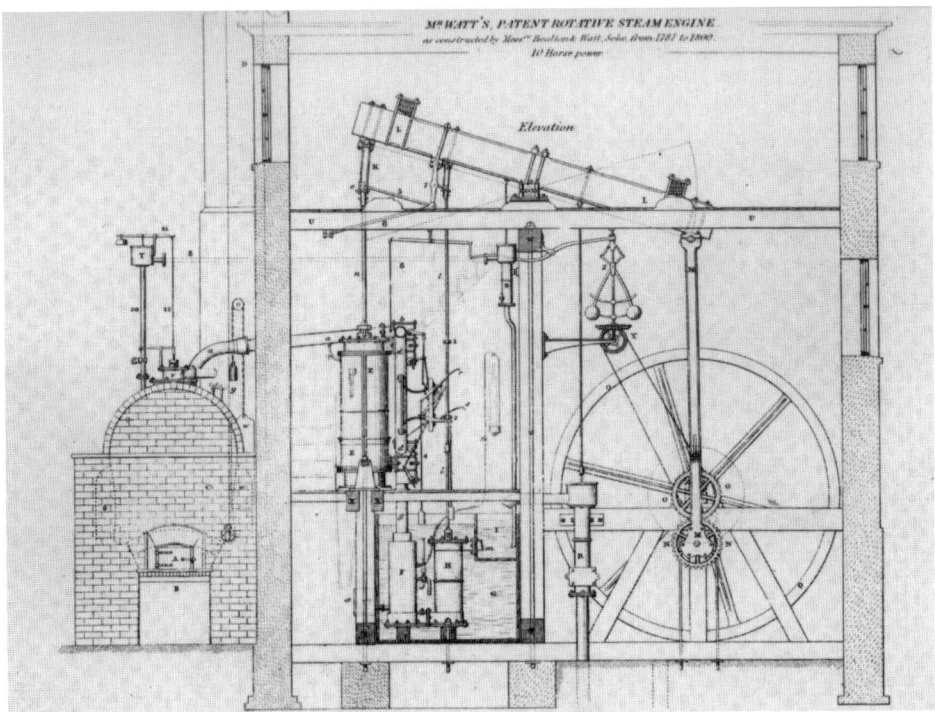

James Watt's rotative beam engine of 1787: one of the first steam engines capable of powering a mill.

in mean, shabby terraces and around airless courtyards. Whatever the fault of the older country mills – and in terms of harsh discipline they were often among the worst – at least their setting was clean. Leaving work, the operatives breathed fresh air. All that changed in the towns. Smoke rose from the mill chimneys to settle as soot on the houses. Dye works sent out their waste to pollute the rivers. There was no proper drainage, no sewers. Oldham brought prosperity for some, but it did not show itself in the streets of the town. A *Morning Chronicle* journalist described how he saw it in 1849:

> The whole place has a shabby underdone look. The general appearance of the operatives' housing is filthy and smouldering. Airless little back streets and close nasty courts are common: pieces of dismal waste-ground – all covered with wreaths of mud and piles of blackened brick – separate the mills.

The bankers and merchants who formed the majority of the mill owners had their houses built on the outskirts of town, solid stone villas set well apart from the slums developing in the centre. The pattern was changing. At Styal and Cromford, the owners had their houses built to overlook the works. In the new towns, the owners withdrew to their own privileged enclaves, while in the centre, numbers grew as looms were set up wherever space could be found. The Reverend Richard Warner in *A Tour Through the Northern Counties of England* (1802) reported that nowhere was the pressure greater than in the cotton capital, Manchester:

> An idea of the immense population of the country in the environs of Manchester burst upon the mind of a sudden, when we reached the summit of a hill about tow miles without the town, where a prodigious champaign of county, was opened up to us, watered by the Irwell, filled with works of art; mansions, villages, manufactories and that gigantic parent of the whole, the widely-spreading town of Manchester.

Manchester was both the mercantile and industrial centre of Lancashire. It was important even before the industrial revolution wrought its changes. At the beginning of the 1790s, for example, Manchester's trade directory listed eighty-seven fustian manufacturers and fifty check manufacturers, thirty printers and dyers, eighteen yarn merchants and six cotton merchants. Altogether, there were some 1,400 entries. To move on 100 years is to see an astonishing change.

The 1,400 entries had grown to nearly 30,000 and there were some 500 cotton manufacturers, more than 400 printers, a whole new group of engineering firms, making machinery for the industries, and scores of cotton merchants, agents and brokers. In the late 1780s, the population was estimated at around 40,000; fifty years later it had more than trebled. Many of the new workers came over from Ireland or from the poorer parts of mainland Britain, but all were drawn by the lure of a boom-time. Handloom

Mills along the Irwell: squalid housing was squeezed in between the tall factories.

weavers' wages rocketed, though the European wars sent them plunging down again just as quickly. But for a time, the weavers were the new elite. The problem, however, had still to be solved – where were they all to live? The answer seemed to be – insanitary slums.

John Ferriar, physician to the Manchester Infirmary at the end of the eighteenth century, wrote a dispassionate account in a pamphlet of the conditions he discovered. His matter-of-fact tone makes it seem all the more chilling. He offered

a few observations on the means of opposing the production and progress of infectious fevers, in cellar and lodging houses, where they reduce great numbers of the Industrial Poor to extreme distress, and often nearly destroy whole families.

It is worth quoting at some length for it is the accumulation of Ferriar's drily noted facts that makes the pamphlet so compelling.

1. In some parts of the town, cellars are so damp as to be unfit for habitation … I have known several industrious families lost to the Community, by a short residence in damp cellars
2. The Poor often suffer much from the shattered state of cellar windows … the consequences to the inhabitants are of the most serious kind. Fevers are among the usual effects.
3. I am persuaded that mischief frequently arises, from a practice common in many back streets of leaving the vaults of privies open … fevers prevail most in houses exposed to the effluvia of dunghills in such situations.
4. In Blakeley-street, under No. 4, is a range of cellars, let out to lodgers … They consist of four rooms communicating with each other, of which the two centre rooms are completely dark; the fourth is very ill lighted, and chiefly ventilated thro' the others. They contain from four to five beds in each, and are already extremely dirty.
5. The lodging houses, near the extremities of the town, produce many fevers … The most fatal consequences have resulted from a nest of lodging houses in Brook's entry … in these houses, a very dangerous fever constantly subsists, and has subsisted for a considerable number of years. I have known nine patients confined in fevers at the same time, in one of these houses, and crammed into three small dirty rooms … Four of these poor creatures died, absolutely from the want of common offices of humanity … The horror of these houses cannot easily be descried: a lodger fresh from the country often lies down in a bed, filled with infection from the last tenant, or from which the corpse of a victim to fever has only been removed a few hours before.

On he goes through a catalogue of horrors. Ferriar continued his investigations, and thirteen years later found things were no better, as he reported to the Manchester Board of Health. Slum houses were being built with windows that were never intended to be opened and even the supposedly well-off operatives of the mighty McConnel and Kennedy, the most powerful cotton manufacturers, lived in back-to-backs with no water supply, no yards or gardens and no paved roads. Pamphlets were written, but no one it seemed heeded the warnings. Fever struck in 1796 and again in 1832, when cholera spread through large parts of the town. John Fielden, MP for Oldham, and himself a manufacturer at Todmorden, rounded on his fellows in *The Curse of the Factory System* (1836). Here was no calm analysis but an impassioned outburst:

> We were panic-stricken, we knew our sins, we recollected the fevers of 1796; we knew the 'squalid homes' of those who made our wealth, we knew the malignancy would fix on them, and that this would endanger ours. It was there that we flew, not from charitable motives, but to save ourselves: we visited, we scoured, we whitewashed (would to God we could whitewash ourselves) we did all that men could do – to save our own! We found hunger, nakedness, bare earthen floors and unfurnished houses with unwhited walls; what of that! Neither of these was catching. Ah! But the pestilence was! We sought out pestilence where we were sure to find it: we did not carry charity where we always knew it was wanted. We were moved exactly as we were in 1796, not by the love of our neighbours, but by fear of the visitations of God.

Of all those who came to investigate conditions in Manchester in the first half of the nineteenth century, none gave a more detailed description, nor more vehemently denounced what he saw, than the 24-year-old son of a German cotton manufacturer, Frederick Engels. *The Condition of the Working Class in England* is one of the key works of Marxist literature and, probably for that reason, has no shortage of either critics or defendants. One has to be wary, however, of putting the Marxist cart before the Engels horse. He did not set out to blacken Manchester to bolster up a communist doctrine – rather the doctrine grew from what he had seen in the town. He obtained much of his material from direct observation and a good deal from Mary Burns, the Irish mill girl who lived with him as his wife until her death in 1863. From his descriptions of walks round the poorest districts, it is clear that what he saw disgusted him and there is little doubt that he described what he saw as accurately as he could. Here he is describing the River Irk and its surroundings:

> At the bottom flows, or rather stagnates, the Irk, a narrow coal-black, foul-smelling stream, full of debris and refuse, which it deposits on the shallower right bank. In dry weather, a long string of the most disgusting, blackish-green, slime pools are left standing on this bank, from the depths of which bubbles of miasmic gas constantly arise and give forth a stench unendurable even on the bridge forty or fifty feet above the surface of the stream. But besides this, the stream

itself is checked every few paces by high weirs, behind which slime and refuse accumulate and rot in thick masses. Above the bridge are tanneries, bone mills, and gasworks, from which all drains and refuse find their way into the Irk, which receives further the contents of all the neighbouring sewers and privies. It may be easily imagined, therefore, what sort of residue the stream deposits. Below the bridge, you look upon the piles of debris, the refuse, filth and offal from the courts on the steep left bank; here each house is packed close behind its neighbour and a piece of each is visible, all black, smoky, crumbling, ancient with broken panes and window-frames. The background is furnished by old barrack-like factory buildings …

Above Ducie Bridge, the left bank grows more flat and the right bank steeper, but the conditions of the dwellings on both banks grows worse rather than better. He who turns to the left here from the main street, Long Millgate, is lost: he wanders from one court to another, turns countless corners, passes nothing but narrow, filthy nooks and alleys, until after a few minutes he has lost all clue, and knows not whither to turn. Everywhere half or wholly ruined buildings, some of them actually uninhabited which means a good deal here; rarely wooden or stone floor to be seen in the houses, almost uniformly broken, ill-fitting windows and doors, and a state of filth! Everywhere heaps of debris, refuse, and offal; standing pools and gutter, and a stench which alone would make it impossible for a human being to live in such a district.

Engels dwelt on one side of Manchester life, others looked and saw only a pathway to a glorious future. They saw the town's businessmen as leaders of a movement that was dragging Britain from a static past to a dynamic future. The great mills and warehouses stood as proud as stately homes. The homes of the manufacturers were the epitome of solid respectability. It was a place of contrasts, but whatever the filth spawned in the new slums; however much dirt might drift and fall from factory chimneys, the manufacturers could always point to the statistics to define progress. Cotton manufacture went up and up in value – £1 million a year in the 1780s, £6 million by 1800, over £20 million in 1820 and £30 million by 1830. And all the time, prices were falling: there seemed no reason why it should not go one forever.

Yarn was exported around the world and there never seemed to be enough weavers to handle the outflow from the mills. The next step was inevitable; just as hand spinning had been replaced by machinery, now it was time to mechanise weaving. The answer had, in fact, been found quite early in the history of mechanisation, and the man responsible was the least likely of all our inventors, the Rector of Goadby Marwood in Leicestershire, Dr Edmund Cartwright. The circumstances surrounding the invention are even more unlikely. The story goes that Cartwright was at a dinner party at which one of the guests remarked

The mills along the Irwell might have been gloomy, but they brought a new prosperity to Manchester and such fine buildings as the Exchange.

that no one had ever invented a power loom, and no one ever would. Cartwright rose to the challenge. It was said that he had never actually seen a loom at work, and he had as little knowledge of the fundamentals of weaving when he finished his product as he had when he started.

His first effort at a new loom was powered by two local men, who turned the handle that drove the mechanism for two hours, before collapsing exhausted. It was a crude device, but a point had been made – the thing was possible. Cloth had been woven on a machine without the help of a skilled weaver. It was a very odd machine. The warp was held in a vertical frame, and the shuttle was sent across by a spring-loaded device which, in the inventor's own words was 'strong enough to have thrown a Congreve rocket'. Odd it might have been, but it worked well enough for Cartwright to start taking the project more seriously. He decided it was time to study a conventional loom, and after that began to work on developing a practical power loom. In 1786 he took out a patent and began to ruminate on the pleasant prospect of profits. He went to Manchester to try and arouse the interest of the local manufacturers, but met with less enthusiasm, so he set up on his own. He bought a mill at Doncaster and began building looms. He was generally mocked, and few were prepared to take a weaving vicar – and a poetic one at that – seriously.

Mr. Cartwright was once Professor of Poetry at Oxford, & really was a good Poet himself – But it seems he has left the Barren Mountains of Parnassus & the fountain of Helicon for other mountains and other vales & streams of Yorkshire, & he has left them, to work on the wild Large & Open Fields of Mechanics … you say not a word about the probability of success, likely to attend his weaving invention so confidently & so flatteringly held out to the world? Do let us have your most candid opinion & distinguish between what is Visionary & what may be practicable in the is new Machaine Machine.

His new versions were very different from the first experimental model, very similar in appearance to a handloom, with the warp held horizontally. Cams were used in place of the operatives' feet to raise and lower the headless and another system of levers and cams was used to work pickers that threw the shuttle. In order to make the machine profitable, it was essential that the operative would need to tend more than one loom, so there was another problem to overcome. If threads broke or the shuttle jammed, then the machine would have to be stopped or the cloth would be ruined. A handloom weaver would at once be aware what was wrong and could act at once. Cartwright had to devise a method of stopping the loom automatically and he invented what was called the warp-protector mechanism. He was soon setting his looms to work, with one operative for every two looms. A number of people came to see this new wonder of technology, including the inventor's friend, the poet George Crabbe, who brought his wife along to share the experience. Her son described her reaction.

When she entered the building, full of engines thundering with relentless power, yet under the apparent management of children, the bare idea of the inevitable hazards on such stupendous undertakings quite overcame her feelings and she burst into tears.

Others were less sensitive and the sight of the factory at work inspired one Manchester manufacturer to begin work on a new mill, with 500 looms, powered by a steam engine. No more than twenty looms had been installed when the building was mysteriously burned down. Was it accident or arson? Nobody knew, but local manufacturers had their own ideas. There was no great stampede to install power looms. In 1793, Cartwright's mill closed down, leaving the inventor heavily in debt. Perhaps it was down to faults in the early machines, or it may have been down to the fact that Cartwright, like many inventors, lacked the business acumen to make the enterprise profitable. But the machines did work, and it was only a matter of time before their use would become widespread.

In 1770, every part of the cotton manufacturing process had been carried out by hand workers, many of them working in their own homes. By the end of the century, it was possible to transfer all these processes from home to factory,

Power looms driven by steam.

powered by water wheel or steam. The transformation was astonishing but not necessarily universally welcomed. But there was no denying the huge increase in production. Edward Baines, looking back in 1835, perfectly expressed the sense of awe and wonder felt by many who viewed the industrial scene:

> It is by iron fingers, teeth, and wheels, moving with exhaustless energy and devouring speed, that the cotton is opened, cleaned, spread, carded, drawn, roved, spun, wound, warped, dressed and woven … Men in the mean while, have merely to attend on this wonderful series of mechanism, to supply it with work, to oil its joints, and to check its slight and infrequent irregularities; each workman performing or rather superintending, as much work as could have been done by *two or three hundred men* sixty years ago. At the approach of darkness the building is illuminated with jets of flame, whose brilliance mimics the light of day …. When it is remembered that all these inventions have been made within the last seventy years, it must be acknowledged that the cotton mill presents the most striking example of the dominion obtained by human science over the powers of nature, of which modern times can boast.

Statistics don't often make exciting reading but those for the cotton trade tell a dramatic story. In 1787 there were 119 cotton mills in Great Britain. Half a century later, there were 1,791 providing employment for over half a million people. More mills devoured more cotton and what was true of Manchester was true of the cotton districts as a whole. The first part of the nineteenth century saw the consumption of cotton in British mills doubling almost every decade: 56 million lbs in 1800; 123 million in 1810; 152 million in 1820; 263 million in 1830 and

Lancashire loom
shed: Roe-Lee Mils,
Blackburn in 1905.

572 million in 1840. As the Empire expanded, so new markets opened up. As an American commentator put it, 'There is not a battle that England has fought in India, Afghanistan or China … that did not extend the consumption of cotton.'

By the middle of the century, the pattern of the cotton boom was fairly set. Each passing year saw the British manufacturers increasing their dependence on America as the source of their raw material. By the 1840s, over 80 per cent of the cotton reaching British mills came for the American South. In the 1790s, cotton production had been limited to South Carolina and Georgia, but by the 1830s it had reached nine states, spreading as far west as Arkansas. During that time, annual production rose from 2 million to 600 million lbs. Not all that cotton was bound for British mills.

Chapter 9

PAWTUCKET AND AFTER

The British having established a lead in the new factory age, were, not surprisingly, keen to hold on to it. They passed legislation banning the export of textile machinery and models and designs for such machinery and, to make doubly sure secrecy was maintained, they also banned anyone who had worked in a mill from emigrating. Secrets, as so many inventors had discovered, are not so easily kept, especially when there were gentlemen across the Atlantic prepared to pay hard cash to anyone prepared to reveal them. Industrial espionage is not a modern concept. A Philadelphian by the name of Tench Coxe persuaded English workmen to make a model of an Arkwright machine, but it never got past the vigilant British Customs officers. It was to be a very temporary success.

In 1782, a 14-year-old boy, Samuel Slater was one of many youngsters who came to work for the Arkwright-Strutt partnership. Seven years later, hearing of the tempting offers from Philadelphian Society of Artists and Manufacturers, he set out for New York, in defiance of the law. This time no one spotted the departure of the Arkwright apprentice, largely because he carried no documents – all the information he needed was safely tucked away in his own brain. He found New Yorkers to be disappointingly slow in taking up his ideas, so he accepted an invitation from Moses Brown of Providence, Rhode Island to make the trip to New England. There he found Brown, and his son-in-law, William Almy, ready to finance a mill in exchange for the information he had carried across the Atlantic, and he was to receive a half share in the profits. Agreement was reached and a site was selected by the falls on the Pawtucket River, which would secure adequate water supply. Slater lived up to his part of the bargain. Working from memory, he planned the water courses, designed carding engines and built a water frame with twenty-four spindles. It was a small beginning, but everything worked and America's first cotton mill was ready to go into production.

There was, however, one obstacle to success. Arkwright and Strutt had been able to tap into a vast reservoir of poor families for their labour force. America had no similar source of cheap labour. It was a vast country with a small population. Land was there for the taking and there was little shortage of opportunities for those who wanted to make their own way in the world. At first the tiny mill was run using child labour – seven boys and two girls representing the entire workforce. But as the mill prospered it was soon evident that something extra was needed. Slater, having borrowed his notions on mill engineering from Britain, now proceeded to borrow social engineering ideas as well. He built company

Samuel Slater's Pawtucket Mill, the first cotton mill in America.

houses, just as Arkwright had done, offering families a home and guaranteed work for all. It proved to be sufficiently tempting to attract the workforce Slater needed. The New England cotton industry was on its way.

At much the same time, an apparently unconnected event was to lead to the next phase of development. In 1792 an enterprising group in Massachusetts set out a notion for developing a canal that would link the upper reaches of the Merrimack River to the coast. They formed the Proprietors of the Locks and Canals on the Merrimack River and set to work constructing a canal to Newburyport. Unfortunately for them, a second company, the Middlesex Canal Company, began constructing a route from Chelmsford on the Merrimack to Boston. The latter succeeded, and the first company failed. But if it failed in its objective, the effort was by no means wasted, though it was to be some time before the endeavour was to bear fruit.

In Britain, new textile inventions were leading on to ever greater efficiency and the same laws protecting machinery were still in place, long after Slater had slipped through the security net. Very efficient they proved, too, in keeping the physical evidence within the country, but there was no protection against the enquiring mind. Francis Cabot Lowell visited Britain's textile mills and proved to have a memory equal to Slater's, for he carried back in his mind the details of the new power looms. Back home at Boston with the help of a mechanic called Moody, he reconstructed a loom. It was completed in 1814, and Lowell invited a local merchant, Nathan Appleton to inspect it. Appleton recorded his first impression in *Introduction to the Power Loom* (1858). 'I well recollect the state of admiration and satisfaction with which we sat by the hour watching the beautiful movement of the is new and wonderful machine, destined as it evidently was, to change the character of all textile industry.' America's first composite mill, combining spinning and weaving was opened at Waltham. It was a small-scale affair, a pilot scheme that was to lead directly to the development of America's first major mill town.

Lowell himself died young, in 1817, but his brother-in-law, Patrick Jackson, together with Appleton and Moody, continued the work and began looking for a suitable site. It was now that the old canal built above the Pawtucket Falls was remembered and in November 1821, the Bostonians, together with Kitt Boott, who had been appointed to manage the new enterprise, visited the site. They liked what they saw and decided that it was here that a mill would be constructed and that the town to be built around it should be called Lowell, after the man with the photographic memory. At the time of the first visit, one of the group remarked that they might live to see the town reach a population of 20,000. In fact, by the time Appleton died in 1861 it had almost reached almost twice that. The old canal was restored to bring water to the new mill and the system was to be extended over the years into the most sophisticated example of water power to be seen anywhere, with over 5 miles of canal to turn the various wheels and turbines. That was in the future, but on 3 January 1824, Boott was able to record that '10 bales of goods sent off to Boston – being the first lot sent from the Merrimack'. The first but not the last; Lowell grew and grew, so that fifty years after those first bales were sent out, there were 100 mills in the town, consuming 1 million lbs of cotton a week. It was to become known as the Manchester of America. The original mill owners, however, were determined that the new town should only comply in part with its English forerunner. Manchester would be matched in productivity and quality – but not in exploitation and squalor.

The motives of the proprietors were not entirely benevolent. The labour recruitment problems that had faced Slater were no less pressing at Lowell. Here, however, a quite new source of labour was noticed. Throughout the region, were large numbers of 'solitary women' – which mean little more than unmarried young women. They were single and needed money – to save up for a dowry if they were to get married, or as income if they were not. There were few opportunities for them apart from the drudgery of domestic service, the miserable wages of seamstresses and teachers or the depressing prospect of life as the poor relation. Mill work was not especially well paid – the highest wages for women were no more than $4 a week when the mill opened, 50c less than paid to the lowest paid male – but it was still more than was offered by the alternatives. There was, however, something of a problem. The idea of good, decent New England farm girls being sent off alone to the dangers and temptations of an industrial town was by no means popular. It was up to the owners to prove to the parents – and to the girls themselves – that mill work

Mills along the Merrimack at Lowell in the 1830s.

was respectable and that the girls would live in a community of unimpeachable morality. In meeting those requirements, the owners of Lowell set up a system that was to make the town famous, not just in America, but throughout the industrial world.

The mill owners, in effect, guaranteed the moral welfare of their female charges by building boarding houses where they were lodged under the care of landladies whose probity and piety matched the highest New England standards. As each new mill went up, so the boarding houses went up with it. Here the girls were fed, housed – and controlled. A list of boarding house regulations for the 1840s makes it quite clear what was expected of the landladies. They were not just moral guardians; they were moral narks as well. They were 'considered answerable for any improper conduct in their houses' and were to 'report the names of such as are guilty of any improper conduct, or are not in the regular habit of attending public worship'. The houses acquired a high moral tone which the girls themselves were anxious to maintain, for there was still the whiff of wickedness about the name of mill girl. Concern for moral welfare was more assiduously pursued than physical well-being. The regulations also demanded that 'the buildings and yards about them must be kept clean and in good order'. Yet many houses were verminous and unbearably stuffy, with as many as eight people living in one room. And mill conditions were little, if any, better than those in the British mills that were so roundly condemned by so many American commentators. The girls worked twelve to fourteen hours

Lowell boarding houses with the Merrimack Mill at the end of the road.

a day, and by the time money had been deducted for board and lodging, were often left with as little as $1 a week for all their labours.

Yet in spite of the hard work, the long hours and the crowded living accommodation, the view of the mill girls that has come down is of a group which was bright, intelligent, hard-working and, above all, devoted to self-improvement. They clubbed together to buy pianos for the houses, attended lectures and even started their own magazine, *The Lowell Offering*. Everyone who visited Lowell, from Davy Crockett to Charles Dickens, spoke of their liveliness and intelligence, though few also noted that all their efforts produced little material reward and less status. It is not too surprising to find some remarkable radicals rising from their ranks, women such as Harriet Robinson who worked for women's suffrage and to improve the life of the mill girls, and Lucy Larcom who went on from *The Lowell Offering* to become a professional author and now has a park named after her in Lowell.

In a sense, the story of the mill girls is a success story, but one that was to be short-lived. Lowell in time was to follow the path of other industrial concerns across the Atlantic. Work rates were raised, wages lowered and conditions deteriorated to such an extent that the New England girls could no longer be persuaded to leave their old lives to work in the mills. Others were, however, ready to take their place; refugees from the famines in Ireland and political revolutions in Europe. For a while it had seemed that a new kind of industrial society might indeed have been built up on the banks of the Merrimack, but the ideas of the Old

The magazine produced by the mill girls of Lowell.

World were now transplanted to the New. The mill girls, pride of Lowell, were no more. Visitors no longer came to stare at them and admire, as William T. Thompson of the *Savannah Morning News* had done in 1845:

> They cum swarming out of the factories like bees out of a hive, and spreadin in every direction, filled the streets so that nothin else was to be seen but platoons of sun-bonnets, with long capes hangin over the shoulders of the factory gals. Thousands upon thousands of 'em was passin along the streets, all lookin happy and cheerful and neat and clean and butiful, as if they were boarding-school misses just from their books. It was indeed a interesting sight, and a gratifyin one to a person who had always thought that the opparatives as they call 'em in the Northern factories was the most miserable kind of people in the world.

Thompson's previous view of factory workers was one shared by many of his contemporaries in the South. It was one of the factors that prevented the establishment of a major manufacturing element in the region. Logic suggested

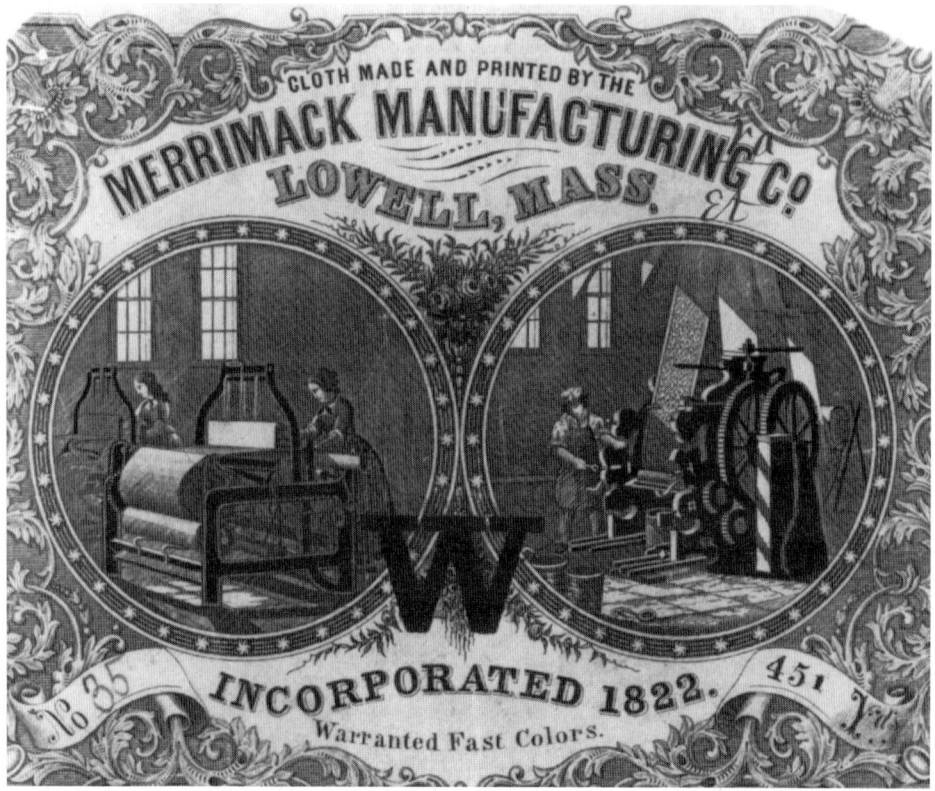

Merrimack company label.

that it made very little sense to pick the cotton in the South, spend a good deal of money sending it North, and then spend even more money bringing finished cloth back South. As early as 1808 there was a notably ambitious scheme begun at Charleston, where the South Carolina Homespun Company was formed largely due to the efforts of John Shecut. He was so confident of success that he named his daughter Carolina Homespun, though what his daughter made of her curious name is not recorded. She was probably less than pleased to find she was lumbered with a name associated with what turned out to be a disastrous flop. This was true of many Southern mills that foundered under bad management. Men who knew nothing of the business hired managers who knew less, simply on the grounds that they claimed to have worked in a mill somewhere or at least seen one somewhere else.

The mills were largely manned by poor whites, but numerous attempts were made to employ black children too young for field work. Governor David Williams of South Carolina was an early factory owner who trained slave children from his plantation and kept them in the mill until they reached an age when they were reckoned to be worth $20 a year as field hands. Then they were sent from the mill and replaced by other 'little hominy eaters' as he called them. But Williams soon found, as others had done, that mill work did not help in the development of good, strong field hands. His human investment was suffering. So, beset by bad management and labour problems, the South struggled to set up an industry beside the plantations. But even as late as 1860, the South employed a mere 8 per cent of the mill workers in America. And the country as a whole showed little sign of catching up on the British lead in manufacturing. With American cotton being exported around the world, more than half the crop was still going to Britain. Some might complain – and many Americans did complain – but that was the indisputable fact.

> There is in reality but one great cotton mill, and that belongs to England, and her agent sits at Liverpool, and sees our labour in bales of cotton, piled up around him till it will cover a ten-acre field. The reports of that market will show a stock, sometimes a million of bales, that stand in Liverpool unsold: with a knowledge of the fact, that it cannot be taken anywhere else. There are the spindles, and there it must stay.

On a world scale, the manufacturing capacity of the South was infinitesimal, but its role as a producer of raw material was vital. For better or worse, the fates of Lancashire and the South were tied together. The destiny of the plantation slave and the mill worker were both determined by decisions taken on the Manchester Exchange. Other factors might have an effect, other countries might be involved, but it was on this unique link across the Atlantic that the world of cotton ultimately depended.

Chapter 10

THE PLANTER

The spread of the factory system in Britain gave rise to a new social order. Once society had been thought of as a pyramid with the royal family at the apex, descending down through the aristocracy to the gentry and on down through tradesmen to the broad base of the poor. It was a system apparently as solid and lasting as the actual pyramids of Egypt. Now, however, people were starting to think of themselves in terms of new relationships: master and worker; middle class and working class; capitalist and labourer – whichever label you preferred these were recognisable divisions. America was too young a country to have established the rigid patterns to match the old hierarchies of Europe. So, the pattern imposed by the spread of the cotton plantations in the South did not seem to involve any dramatic break with the past. Yet a new type of society was evolving and the changes were profound. Where Britain was developing a new class system, the Southern States were developing one based on racial divisions, master and slave. The point is so blindingly obvious that it might seem scarcely worth mentioning at all, but it marks the crucial difference between the experience of industrialisation and the changes on the plantations.

The textile worker in Britain had, in theory at any rate, a prospect of change held out in front of him. This could either be through that beloved doctrine of the Victorian middle class – self-help – or through new mass organisations that would try and act on behalf of the workers. And, however slowly, the position of the textile worker did shift. For the black slave, however, the plantations were part of a totally rigid system. There could be no change in the slave's status or condition. Where the spread of the cotton industry was forcing change in British society; it served only to fix the South in its ways. In place of a doctrine of change, the plantations only offered near stagnation; cultural, social and technological. The Manchester of 1850 was nothing like the town of half a century earlier; that same time span showed no discernible change in plantation life.

The society of the plantation was based on the planter as absolute monarch. There was no such thing as a typical plantation, any more than there was a typical mill, or a typical planter. Yet a stereotype has persisted, bolstered over the years by novels and later the cinema. The image is of a pampered existence of grand houses and Southern hospitality – of gambling and a code of honour among the men, acquiescence in comfortable domesticity for the women. The shock comes in finding how frequently these unlikely fictions turn out to be based on reality.

The wealth of the plantation seen in the Greek revival style: the Colley-Barksdale house in Washington, Georgia.

John Quitman arrived in Natchez in 1821 to practice law and described a society in which the planters ruled as to the manner born:

> In the city proper, and the surrounding country, there is genteel and well-regulated society … The planters are the prominent feature. They ride fine horses, are followed by well-dressed and very aristocratic servants, but affect great simplicity of costume themselves – straw hats and no neckcloths in summer; and in the winter coarse shoes and blanket overcoats. They live profusely; drink costly Port, Madeira, and sherry, after the English fashion, and are exceedingly hospitable.

Life seemed idyllic.

> Mint-juleps in the morning are sent to our room, and then follows a delightful breakfast in the open veranda. We hunt, ride, fish, pay morning visits, play chess, read or lounge until dinner, which is served at two p.m. in great variety, and most delicately cooked in what is here called the Creole style, and many made or mixed dishes. In two hours afterwards every body – white and black – has disappeared – the siesta of the Italians. The ladies retire to their apartments, and the gentlemen on sofas, settees, benches, hammocks, and often gipsy fashion, on the grass under the spreading oaks. Here too, in fine weather, the tea-table is always set before sunset, and then, until bedtime, we stroll, sing, play whist, or croquet. It is an indolent, yet charming life, and one quits thinking and turned to dreaming.

Reading through planters' diaries can often seem like dropping in on a character from a Jane Austen novel. Everard Green Baker notes in his diary that he has read Johnson's life of Dryden and finds him 'a prolific author wielding a pen powerful, elegant, majestic' and resolves 'as part of my daily literary employ' to read all the lives of the poets at the rate of one poet per day. He hunts, visits friends, and, in the evening, when he has finally run through his catalogue of poets, turns with equal pleasure to Macaulay. Ladies are, if anything even more Austenish. Mary Bateman's life, for example, was a round of visits and music lessons, teasing about 'The Judge' and excursions in the family carriage. The young ladies sat listening while gentlemen callers read aloud from Shakespeare, and rather more daringly, Byron. The arrival of a boat on which the ladies could sit for a daguerreotype completely monopolised the diary for three days.

That fine home, however, had to be built in the first place, and even then was not always as fine as tradition would have us believe. Joseph Thompson wrote to his aunt in Alabama from Louisiana, where he had gone to start a plantation:

> I have undergone some hardships since I left you. But notwithstanding my health has been good. I am at this time living in an open Cabin that keeps but little

The Southern planter and his wife.

wind out and it is very chilly cold at this time. We had ice a plenty this morning. The grass has taken a fine start but this change will check .... I found provisions hard to get until my friend Ned arrived since I have faired sumptuously of your kind present the barrel of flour .... I am in a country that is full of sharpers and from all parts. There is about 20 new settlers within the limits of six miles around here and more expected daily ... the face of the country is not so level as I thought it to be last fall and the soil is not so dark as I thought it to be. But they all say that Cotton will grow and that is the main object in this country.

The British actress Fanny Kemble went out to join her husband on a Georgian plantation and the record (*Journal of the Residence in a Georgian Plantation* 1863) of her stay for the years 1838–9 is one of the sharpest, as well as one of the most readable, accounts of plantation life. She has been accused of prejudice and lack of understanding – the fate of all foreigners in the country they dare to criticise – but her comments ring true. She admitted that she did arrive with preconceived ideas. 'Assuredly I *am* going prejudiced against slavery, for I am an Englishwoman, in whom the absence of such a prejudice would be disgraceful.' But, as she also made clear, her wish to think the best of her husband inclined her to look for what was good in his life, not dwell on what was bad. But on her arrival, she found a house that fell along way short of her expectations:

It consists of three small rooms, and three still smaller, which would be more appropriately designated as closets, a wooden recess by way of pantry, and a kitchen detached from the dwelling – a mere wooden outhouse, with no floor but the bare earth … Of our three apartments, one is our sitting, eating and *living room* and is sixteen feet by fifteen. The walls are plastered indeed, but neither painted nor papered: it is divided from our bedroom (a similarly elegant and comfortable chamber) by a dingy wooden partition covered all over by hooks, pegs, and nails, to which hats, caps, keys &c &c are suspended in graceful irregularity. The doors open by wooden latches, raised by small pieces of packthread – I imagine the same primitive order of fastening celebrated in the touching chronicle of Red Riding Hood; how they shut I will not pretend to describe, as the shutting of a door is a process of extremely rare occurrence throughout the whole Southern country. The third room, a chamber with sloping ceiling, immediately over our sitting-room and under the roof, is appropriated to the nurse and my two babies. Of the closets, one is Mr. – the overseer's bedroom, the other his office or place of business; and the third, adjoining our bedroom, and opening immediately out of doors, is Mr. –'s dressing room and cabinet d'affaires, where he gives audience to the negroes, redresses grievances, distributes red woollen caps (a singular gratification to a slave) shaves himself, and performs the other offices of his toilet. Such being our abode, I think you will allow there is little danger of my being dazzled by the luxurious splendour of a Southern slave residence.

The life itself was far from healthy. Any plantation diary can be virtually guaranteed to have descriptions of illnesses, most, not too surprisingly, in the slave quarters. Most of the diseases were associated with poor diet and bad

The planter's first crude home at the Callaway Plantation, Washington, Georgia.

The second-generation house at Callaway, the wooden federal-style house.

sanitation, stomach complaints were commonplace, dysentery and cholera by no means rare. Lucian Polk of Mississippi recorded that he had lost twenty-seven slaves with cholera 'produced it was supposed by eating sour meat and corn' – which, of course, he had himself provided. When an epidemic hit, however, it was no respecter of colour – master and slave could both fall victim. A South Carolina planter wrote that 'out of about fifty souls, white and black on the plantation not one escaped the fever and I lost my lovely daughter Thirza'. But disease hit hardest at the slave children, prey to scarlet fever and pneumonia.

The white masters were also prey to all kinds of medical quackery. There were their own patent remedies which, if they did not cure, occasionally do sound quite appetising – blackcurrant cordial made up of a quart of sweetened juice, to which was added a quart of brandy would probably produce no worse effect than a hangover. It is alarming, however, to read in Everard Baker's diary that that a recipe consisting of a tablespoon of common salt and half a teaspoon of red pepper, is a 'nearly infallible' cure for cholera. Patent medicines were widely peddled, some of them so downright harmful that they probably saw off more than they cured.

If the plantation home was not always a fine Colonial-style mansion, and if the happy children playing among the magnolia trees were liable to be struck down by disease, other aspects of the stereotype do seem to hold good. The gambler and

The author and his wife at a house slave's cabin, Boone Hill Plantation, near Charleston.

the man of honour existed, if in less glamorous forms than the romances suggest. John Nevitt of Mississippi seems the very prototype of the gambler. His diary is full of accounts of idle days and nights at the card table – and of his lack of success. 'Rode to Natchez done nothing set up at Dr. Gustins playing Eucre lost 70 dolls.' In May 1827, he had six big card games. His best day was the twenty-first when he was able to record he had lost nothing. On the other five days he lost amounts varying from $50 to $150 – a grand total of $435 for the month. In December the following year he went to the theatre to see *The Gambler's Fate* and was, he reported, 'greatly affected'. The effect lasted right through to the next day when he was off to the races, where his view of gambling was much improved by his $30 winnings. Nevitt's life, when he was not demonstrating his lack of skill at the card table, was mainly taken up with immensely complex financial transactions. Running the plantation involved a complicated system of debts and credits. Just one entry for one day gives some idea of how tangled affairs could become:

> Gave Edwd Turner a Mortgage on 75 head cattle 30 horses 5 waggons 2 carts and for his endorsement to my note for 1500 Dolls in two several notes one for 1000 Dolls the other 500 Dolls gave Jas. C. Wilkins the one of 500 became due on the 15th Decr next for his acceptance for the same amount disposed of that acceptance to Wm Brun at the rate of 10 per cent per annum paid my acct. to

Brunamt 143 doll purchased meat paid Blackman 25 Dolls to my acct of my due bill which he had lost and gave him my due bill for balance 27.25.

The property was frequently mortgaged and in spite of growing debt, credit seemed always to be available somewhere. Notes were handed out to creditors, payable at some specified future date and a good financial manager could keep all his debts and credits permanently on the move, like a juggler keeping Indian clubs up in the air. But even the best jugglers drop a club sometimes. Nevitt dropped his in January 1831. His debts had caught up with him and he had to sell off part of his estate, together with nineteen slaves to pay off over $10,000. The whole system was summed up Baker, living in similar style:

I am considerably in debt, & it troubles me to think so for I do not feel as free as 'tis pleasant to feel …this credit system is so fascinating so inclined to make a man go farther than prudence would dictate that I do here declare that I will not go in debt for anything more this year, except barely for articles to keep my plantation going.

Gambling, whether at the card table or in terms of selling crops that had yet to be planted let alone harvested, was very much a feature of plantation life. So too was the concept of honour and the duel, though in reality it was often very little better than an undignified brawl. The events described (in *Plantation Life in the Florida Parishes of Louisiana, 1836–46*) here took pace in Louisiana in 1839.

Leigh of Va. Challenged Davis for no cause whatever – Moore took the challenge – Davis told him he would answer him in the morning. Moore then observed it was cowardly Evasion – Davis then struck with his whip – breaking his nose etc. Somewhat of general fight took place – Davis supposing it was all over retired to the Post office – having found that Leigh intended to kill him armed himself – in a few minutes saw Leigh coming towards him in a great hurry – he told him to stop or he would shoot him – Leigh rushed at him with a sword cane drawn – Davis snapped both caps of his gun – then with it knocked Leigh down and could have killed him supposing he had injured him very much threw down his gun and started off – but had taken a few steps before Leigh was at him again – he then retreated 'till he got his pistol out which was entangled in his breast whirled & fired & hit Leigh in the small of the Back – Leigh had fallen in pursuit of Davis but was getting up hen Davis fired – he is expected to die every day, perfectly dead from the wound downwards – Davis is justified in every act.

That was not the end of the affair; now it was Moore who challenged Davis to a duel in revenge for the injury to Leigh. They met at 50 yards with rifles, and Moore was killed by the first shot. Superficially, the planter's life might have seemed one of gentility, hospitality and honourable behaviour, but behind the façade there was a world of violence, disease and debt. The mythology of the South has been stressed, because one part of the myth seems to have been accepted by the planters as being unmistakably true. They had a set of beliefs about the relationship between the masters and the house servants and then acted as if they were facts. And in some cases, it had an element of truth.

The house slaves necessarily had a special relationship with their white masters, very different from that of the field hands. They lived in the same house or very close to it. They were intimately concerned with the family, and, when it came to bringing up the young children, often more involved than the white parents. In the scenario, the loyal, devoted slave was rewarded with the affection of the family. Mammy Harriet, a slave on the Dabney plantation in Mississippi, recalled this scene of a mistress rushing to see a dying slave:

Missis put on her bonnet an' went to her jes' as fast as she could. When grannie saw her she could not speak, but held out both arms to her. Missis run into her arms an' bust out cryin'. She put her arms roun' grannie's neck an' grannie could not speak, but de big tears roll down her cheeks. An' so she die.

House slaves for sale in New Orleans.

Affection was possible and Everard Baker wrote an obituary in his diary for the slave, Jack, who died aged about 65. It is the more moving as it expresses personal views written in a private diary:

> He has been in my mother's family since he was quite small, served my mother faithfully through her life time, & stood high in her regard. Since I have owned him he has been true to me in all respects – he was an obedient servant to his master & mistress, an affectionate husband & father to his family – I have never known him to steal nor lie & he ever set a moral & industrious example to those around him – altho' he was not a professing Christian – yet no man white or black that I have ever known was more exemplary in his conduct.

Baker seems to have been aware that there was something odd in this eulogy but could only conclude, rather lamely, 'He deserves a better reward than can be given in this world'.

While the miseries of slavery are all too evident, it is not always quite so easy to see the damaging effect of slave ownership. Fanny Kemble, coming to the system as an outsider, was able to take a detached view, in her *Journal of a Residence in a Georgian Plantation, 1838–9*. She found the slaves almost overwhelming her with offers of help and watched with alarm as the same thing happened to her young daughter. She noted:

> The universal eagerness with which they sprang to obey her little gestures of command. She said something about a swing, and in less than five minutes headman Frank had erected it for her, and a dozen young slaves were ready to swing little 'missis' … think of learning to rule despotically your fellow creatures before the first lesson of self-government has been well spelt over! It makes me tremble: but I shall find a remedy or remove myself and the child from the misery and ruin.

The house slaves were a group apart, with their own privileges. In the cities they dressed in a style that astonished Northern visitors:

> To see slaves with broadcloth suits, well-fitted and nicely ironed shirts, polished boots, gloves, umbrellas for sunshades, the best of hats, the young men with their blue coats and bright buttons, in the latest style, white Marseilles vests, white pantaloons, brooches in their shirt-bosoms, gold chains, elegant sticks and some old men leaning on their ivory and silver headed staves, as respectable in their attire as any who that day went to the House of God, was more than I was prepared to see.

Not surprisingly, such slaves tended to consider themselves as a cut above the rest.

On hand every day, ready and eager to do just what master or mistress demanded, loyal and faithful towards Massa's family – that was how the planter wished to see his house slaves. The Civil War brought home a new reality. The South discovered that the drama they had written in which the slaves played their parts was, in reality, a farce. The actors stopped playing. To the whites it was a great betrayal: to the blacks the end of a long pretence.

Edwina and Bertha Burnley looked back with sentimental affection when they remembered many of their old slaves, such as Aunt Dicey:

All the little negroes called her 'Ga Muh'. She took charge of all the babies while their mothers were in the field – each baby had its own cradle with an older child to rock and amuse it. In summer the cradles were under a great spreading oak and scores of children under twelve years playing around them. Some of the little ones lay asleep on the grass and Aunt Dicey would caution us not to step over them – because they wouldn't grow any more.

And there was Uncle Banks, the carpenter who built the cotton gin. When he was ill 'pa used to see him every day and talked with him. They were friends'. And then came the change and the trauma:

In the first year after the war Pa was sick a good deal, we had no overseer, the negroes got their provisions regularly but worked when they pleased. One morning Cousin Hexzy rode over to inquire about Pa and found us without a stock of wood, the house servants all gone, and me trying to break up pickets with an axe. He took his gun, went to the quarter and ordered out negroes, set them to work cutting and hauling wood and did not leave until there was a winter's supply piled up at the house. I need hardly say that that year's crop was a failure.

It had all been a sham. The planters had wanted the impossible – total obedience and affection. The obedience had been there while the slaves had no option; the affection was all in their imagination. They had invented a character called the faithful slave and if he existed in reality then that was the justification for handing out the harshest punishment to the unfaithful. The same device was used in the factories of Britain to justify the suppression of discontent. They invented the contented workman:

I live in a cottage and yonder it stands;
And while I can work with these two honest hands,

I'm happy as they that have houses and lands,
Which nobody can deny.

I keep to my workshop all the day long
I sing and I whistle, and this is my song –
'Thank God, which has made me so lusty and strong',
Which nobody can deny.

The author was no lusty workman, but Mr John Byrom MA. Discontent could not just be wished away or laid at the door of outside agitators. This was as true of the mill workers as it was of the slaves in the cotton fields.

Chapter 11

COTTON FOR LANCASHIRE

How ever you view the way of life of the planter, it was ultimately dependent on the crops in the fields. Calculation of profit and loss was even more difficult than it was for the mill owner. The latter bought in raw materials and sold finished goods. At its simplest, the difference between the two prices, once you have costed in labour and depreciation, is the profit on the capital invested. The same calculations for the plantation look very different. You could not treat slaves as wage labourers for they were the property of the slave holder; nor could you treat them simply as property, in the way the mill owner could treat his machines, for machines are inanimate. And machines did not reproduce themselves. Looked at in strictly economic terms, a male and female slave who had children were increasing the value of the owner's property for the children were, like their parents, now part of that property. The owners were very aware of this as plantation rules clearly show. 'Marriage is to be encouraged as it adds to the comfort, happiness and health of those entering upon it, besides ensuring a greater increase. No negro man can have a wife, nor a woman a husband, not belonging to the master.' There were other arguments against allowing outside marriages:

No rule that I have stated is of more importance than that relating to negroes marrying out of the plantation it seems to me, from what observations I have made it is utterly impossible to have any method, or regularity when the men and women are permitted to take wives and husbands indiscriminately, and without being able to assign any good reason, though the motive can be readily perceived, and it is a strong one with them, but one that tends not in the least to the benefit of the Master, or their ultimate good, the inconvenience that at once strike one as arising out of such a practice are these –

First – in allowing the men to marry out of the plantation, you give them an ungovernable right to be frequently absent.

2d – Wherever their wives live, there they consider their homes, consequently they are indifferent to the interest of the plantation to which they actually belong

3d – It creates a feeling of independence, from being, of right, out of the control of the masters for a time –

4th – They are repeatedly exposed to temptations from meeting and associating with negroes from different directions, and with various habits & voices.

Loading cotton for Liverpool in New Orleans.

It seems never to have occurred to the writer that they might have the possibility of meeting some with better habits and greater virtues. There might seem to be many arguments against marrying out, but the slaves were no more in control of their affections than any of the rest of us. They would wish to marry where their love lay, but if they did, they could create a terrible dilemma for themselves. The threat of forced separation always hung over them. The more fortunate might be able to get together if their owners came to terms:

> Dear Sir
> Your negro man Sam states that you wish to know if my Mother will sell his wife (Matilda) and child. She could better spare any other negro that she had than Matilda, and does not wish to sell at any price. But Matilda seems as if she wished to go with her husband and under such circumstances she will take sixteen hundred dollars for her and her child.

A plantation rule book reads like an instruction manual for a new engine. It is all quite exact and precise and not the least prone to variation. The Highland Plantation rule book puts it quite clearly: 'A plantation might be considered as a piece of machinery, to operate successfully, all of its parts should be uniform and exact, and the impelling force regular and steady.' But it was not a machine, and all the rule writing in the world would not make it one, whatever theory might say. And theory said that slaves were merely property. 'I have ever maintained this doctrine', declared David Barrow, a planter from Georgia, 'that my negroes have no time whatever, that they are always liable to my call without questioning for a moment the propriety of it'. But that didn't work either, for however much the owners might deny the humanity of their slaves, that humanity continually asserted itself.

Was the slave system essential for a profitable cotton plantation? There is a certain similarity here between the slave system and the apprentice system in the mills. In the latter case, the majority of mill owners decided that they were better off hiring labour for wages than having all the expense and bother of being responsible for the lives of their apprentices. Might the same sort of thing be true of slave labour? The question is unanswerable as the whole economy of a big plantation was tied as much to the cash value of the slaves as it was to the value of the crops. What really matters, however, is the unshakable belief of the planters that without slaves the plantations would collapse into ruin. All their efforts were turned towards making the slave system work – or in paying other to do that for them.

The 'Peculiar Institution' gave the cotton plantation its unique character, but the planter still faced the same problems as any other farmer. He depended on the health of the crop, the vagaries of the weather and the state of the market. The absentee planters, and there were many of them, relied on regular reports from their overseers to keep them in touch with day-to-day business. Many of these reports were barely literate but they provide the best record we have of how a plantation was run:

An overseer watching the cotton pickers.

> I have the promisingest Crop that i have had since i have been in the miss i will finish in a few ours going over the cotton for the first time my negroes and mules is all fat an you think you go a hed and I say go a hed and a good Crop is the object.
>
> Thir is a lot of negros to be sold at Coffville the first of August and for cash and I expect will be barganes to be bought.

Most owners made regular visits to their plantations, and when things were going well they were cock-a-hoop as is in this letter from Archibald Arrington to his wife:

> I came here on Wednesday last and I found my negroes all well and every thing seemed to be getting along well and prospering – negroes all looking well satisfied pleased with the overseer and have made me an excellent crop of Corn and Cotton – fattened and killed seven thousand pounds of Pork & made me between two hundred and sixty and 270 bales of Cotton weighing 500 pounds besides Some of the women have had children since we left & they are all living and doing well so if I think this is not beating Edgecombe it is doing well enough to satisfy a moderate appetite.

But bad news seemed to keep pace with good: 'We had the hardest rain Tuesday that I ever saw fallen the Rains have injured our cotton in this section powerful.' Some had far worse stories to tell:

> Nothing good to write about … the grass caterpillars that were eating the young corn and grass when you were here have eat up all the grass and commenced eating the corn and the cotton. They are tearing the cotton and corn all to pieces and cane and peas and in fact everything that they can crall on they eat off … I think they are going away now and I hope in a few days they will all be done. They have destroyed a Great Deal of cotton since they commenced. Though they have served us bad enough we have not suffered no loss compared with the people on Flint river below Bainbridge and some above.

A story of ruined crops that could be repeated a thousand times, and all too familiar to any farmer. Equally familiar would be uncertainty about crop prices. The cost of growing cotton varied little from year to year, but the price the planter got for his crop could fluctuate wildly. There was, to some extent, a natural evening-out; prices tended to fall in a bumper year and rise in a bad one. But the American grower faced another problem in that prices were being set thousands of miles away by the cotton merchants of Manchester and Liverpool. The grower did have a number of options open to him; he could deal directly

A Mississippi plantation with a steam powered gin in the background.

with a British agent or employ a factor at a major port such as New Orleans or Savannah. He could even ignore the British market altogether and try to sell to the manufacturers in the North. But ultimately, as the major user, Britain set the price. Most planters were more than happy to leave the problems of the market place to agents and factors.

The Minor family who had a plantation near Natchez, Mississippi, received a steady flow of letters from their factor, giving the current state of the market. One difficulty they constantly faced was the time lag between the cotton leaving the plantation and its arrival in Liverpool. The terms might have been favourable when it started on the journey but rather less so by the time it had completed its Atlantic crossing as is clear from this letter from Barclay Salkeld in May 1819:

> The present vessel has been long delayed by head winds, but as I think there appears some chance of a change I think it necessary to advise you that several vessels are in from Savannah whose arrival with advices of the decline in price in America has damaged our markets & American cottons are down full a halfpenny.

The rivers were the main transport routes for the South. Here the stern-wheel steamer *America* has been loaded with cotton bales.

Each month, the importers sent out a printed bulletin giving current prices and other trading details, and then added a handwritten note to suit a particular client. A bulletin of January 1823 reviewed progress for the previous year:

> That the demand for the Raw material will go on increasing, while prices remain moderate, we have not a doubt; because the low price at which the manufactured article is brought into the market induces an increased consumption at home, and a more extensive demand for foreign countries; which will be aided by the new markets that have recently been opened to us in South America. In order to meet this demand, new mills are erecting with great rapidity.

While the Southern planter viewed falling prices with alarm, they were greeted in Liverpool with cheerful sanguinity. The fall in prices, they assured the world, would not encourage the planter to shift to different crops, but would 'have a direct contrary tendency, by stimulating his exertions to extend the cultivation to the utmost of his means in the hope than an increased quantity may compensate for a decreased price.'

This might seem reasonable in terms of the overall view of the cotton market but might be less appealing to the individual planter. To this uncertainty, the planter could add the risks caused by delays, damage, bad packaging and handling and a

Paddle steamers lining the Levee in New Orleans, unloading cotton for shipment to Liverpool.

host of other possible mishaps. In the circumstances, many found it safer to accept a price offered by an American factor and let him worry about fluctuations in the British market. In theory, this then left him free to concentrate on his crops and improving the quality of the yield. In practice, many planters were quite happy to pass on that job as well. They employed overseers and either stayed in the Big House or returned to the towns, which offered more refined pleasures than those available on a remote and unhealthy plantation. The overseer had responsibility for the success or failure of the crops: happiness or misery for the slaves depended on his temper.

The owners were well aware of the importance of such a key figure, and in order to make sure everything was run exactly as they wanted during their frequent absences, they laid down complex sets of rules, governing all aspects of plantation life. An overseer who obeyed the strict letter of the rules would have been a true paragon of all the virtues. He was expected to set a high moral standard, though the regulation that forbade 'swearing, drinking or any immorality' was only to be applied while on the plantation: elsewhere he could, it seems, sin till he bust (see Appendix).

The overseer controlled everything. He decided what work had to be done, who was to do it and set the length of the working day. His duties continued after work was finished, for he was held responsible for the behaviour of the slaves. Within his own domain he was lord and master – and in matters of law, he was judge, jury and executioner. The rule book laid down a full list of punishable offences. Some prohibitions were specific and obvious; running away coming at the top of the list, followed by stealing and drunkenness. Others were concerned with more personal matters, designed to ensure that the slave had little or no say in the regulation of his own affairs. If the overseer deemed the slave's hut to be dirty, the slave was punished. If he failed to get back indoors by the time the horn blew at the end of the day to announce curfew, he was punished. And there were catch-all rules, such as those dealing with neglect of work. But the rule makers had a problem. How do you punish a slave? You cannot fine him for he has no money; you can't imprison him or you lose his labour. You can't even remove privileges where no privileges exist. So, you are left with the whip, but still you must act with some caution. You have no wish to permanently damage your property, yet discipline must be maintained. 'They must be flogged as seldom as possible yet always when necessary … The highest punishment must not exceed fifty lashes in one Day.'

So once again it all comes down to the temperament of the overseer. In some cases, he was a member of the planter's family, and in other he worked under the direct supervision of the owner. But when he was absolute despot, the scope for abuse was enormous. It was in such cases that most of the worst cases of abuse occurred. Yet there were limits set even here to the overseer's freedom of action. Theoretically, all white men were superior to even the best black man,

Plantation owners and overseer watch over the press where cotton is being screwed down into bales.

whose status could never be higher than that of a chattel. In practice, the wish of the owner to see a well-run plantation gave the slave some leverage to use against an unpopular overseer. If things went wrong, the overseer had to have an explanation ready. Even when a slave ran away, the overseer was quick to point out that it wasn't his fault:

> On last munday Gilbert left home and we believe is aiming to git to Dr Caldwell i think you had best come by thir for I have serct the neighbourhood and Cannot hear of him I do not no what took him of unless it was because he had been stealing i have not struck him one lick in a year nor yet threatened him.

Although access to the master was limited, slaves had ways of making their views known. An unpopular overseer might mean poor crops. The overseer could be replaced, but the slaves were permanent fixtures and the wise owner listened to their views. Archibald Arrington visited his Alabama plantation and found the overseer avoiding him: 'If he does not come to me I intend to send for him, for from what my negroes tell me he is a dishonest man.' The chattels had made their point.

The overseer's position was always anachronistic. He was set up as a member of the superior race, yet the slaves knew that he was as dependent on the whim of the planter as they were. His life was different from theirs, but not that different. He could not demean himself by working in the fields, yet he always had to be present when they worked there. Their movements were limited, but so were his and his status could seem little higher than theirs. The overseer's time is paid for by his employer and belongs this employer. It is not right to use that or anything else belonging to the employer, in going about, in visiting, or in entertaining or in any way but for his employer.'

The missives handed down from absentee owner to planter were as curt and high-handed as any that passed from overseer to slave. Some owners piled regulation upon regulation in an attempt to maintain long-distance order. The Hugenin Plantation Book sets off in high style with a statement of principles.

> In the management of my people I would always demand justice and moderation. But at the same time my people must be kept in perfect order, and if necessary should always be corrected to the extent of the fault or crime committed.
>
> I never allow stealing – getting drunk or any unruly or loud conduct such as quarrelling and fighting. Absence from home without permission impertinence in any form or shape as this when allowed is the first step to the disobeying of all the above regulations and finally leads to dissatisfaction. On the contrary I would wish the moral conduct and appearance of my people well attended to – I think that negroes on a plantation should always be made to appear at least decent, especially when there appears to be a disposition to the contrary.

Everything had been laid out to ensure a perfect order, yet the following year, fresh rules had been added and a different tone taken:

> I positively forbid my negroes working out for any person in the neighbourhood. From what I can understand my negroes are very much in the habit of going about the neighbourhood and on Sundays lurking about peoples houses that must also be stopped. My regulations expressly say that there shall be no visiting except by permission from the overseer, who should be as few times from the plantation as possible on Sundays.

New regulations keep turning up alongside complaints about the old ones not being kept.

The overseer is not one of the most loveable characters in history, but his life must often have been a lonely and miserable one, shut away in the plantation with only the slaves for company, with whom he was expressly forbidden to have any social intercourse at all. The system insisted that black and white could never mingle: overseers and slaves had to maintain that difference at all times. That was the theory, but the practice was very different – and, however, brutal the circumstances – both owners and overseers turned to female slaves as women, and on many plantations there were children whose skin colour told of their parentage. Nowhere was the essential ambivalence seen more clearly than in the children of white fathers and black mothers. 'There is a great deal of talk through the Country about abolition &c. Yet the people submit to Amalgamation in its worst form in this Parish. Josias Grey takes his mulatto children to public places &c. and receives similar company from New Orleans, fine carriages & Horses.'

Fanny Kemble was one of those who saw the contradiction very clearly.

> Now it appears that there is no law in the white man's nature which prevents him from making a coloured woman the mother of his children, but there is a law on his statute books forbidding him to make her his wife; and if we are to admit the theory that the mixing of the races is a monstrosity, it seems almost as curious that the law should prevent him marrying women towards whom they have a natural repugnance, as that education should by law be prohibited to creatures incapable of receiving it.

The evidence of racial mixing was there for all to see throughout the South, visible evidence that when law making and desires clashed, it was often the law that gave way. But none of this meant that the fundamental relationship between white and black had changed. Where did the children stand – did they belong with the white father or the black mother, since the law kept the two apart. The answer was usually with the mother, and their status had not changed. Some claimed that those with lighter skins got preferential treatment, but there is absolutely no evidence to support this. One mixed race slave put the situation all too clearly, pointing out that slavery is bad enough, but it is even worse when the suffering is imposed by your own father.

The overseer has always been seen as one of the villains of the cotton story, but he could also be seen as one of the victims, frequently treated badly by his employer. James Barth applied for an overseer's job in Georgia, writing from a lonely plantation in Alabama:

> My main cause for coming back to live for one or 2 years is I am yet a single man and do not wish to live single any longer and rather come back to old Georgia to get me a wife and I have another cause for quitting is my employer is sutch a hard man to get money out of he has not paid me one Dollar in 3 years.

No dispassionate observer looking at this cumbersome, unwieldy system could call it either sensible or practical, regardless of the underlying morality. Yet, once begun, the plantation system developed a momentum of its own that carried it over the reefs of absurdity. As the demand for cotton grew, so the system grew with it. In Britain, the spread of cotton mills was part of a dynamic process. Old machines were being improved, new ones developed, changing the nature of the work. New markets were being won, a new class system was developing. None of this happened in the South. A slave worked in 1850 just as he would have done half a century earlier. Apart from the gin, no new machines were introduced to speed the labour – ploughing, sowing, weeding, harvesting – all went on just as they had before.

Slavery was a disincentive to growth: the absentee owner counted his wealth as much in the human flesh he owned as he did in the crop he sold. It would not go on forever, and the poor agricultural practices were beginning to have an effect, clearly shown in this letter from Hannah Hinman Day to John P. Broune in April 1842:

> Southern planters are getting poorer every year now, this is acknowledged by all. They own large tracks of land which they plant in one patch till it is worn out, then another, till all is worn out, never cultivate the soil and pay enormous tax for labor in supporting their slaves, which is another cause of their depreciation; crops never will again bring former prices and if I could (said he) sell my land for any decent price I would transfer my property to the North. Tell your son-in-law I advise him for the time is short to the abolition of slavery. Freedom is now spreading here and soon will soar and plant her standard throughout the globe. The people of the earth are now becoming a thinking and inquiring race. Look at chartism in England, look at the King of the French, encased in armour, defended by a numerous armed guard.

The author of the above quote was no abolitionist, but, in his own words, 'almost a hater of negroes'. But was freedom really spreading its wings in the South or over the mills of Britain?

Chapter 12

SUCCESS TO THE RISING OF THE WHITES

The chapter title is taken from John Wade's *Extraordinary Black Book*. (1831) in which he put forward the views of a rather restrained form of radicalism.

> We are not of the number of those who inculcate patient submission to undeserved oppression. A favourite toast of Dr. Johnson's was, 'Success to the insurrection of the Blacks'. Shall we not say – Success to the Rising of the Whites! We should at once answer yes, did we not think some measures would be speedily adopted to mitigate the bitter privations and avert the further degradation of the labouring classes.

Insurrection for the blacks, constitutional reform for the whites: many who viewed the scene on both sides of the Atlantic would have thought the reverse would be more likely. The British would follow the French revolutionary path, while the chances of blacks violently taking matters into their own hands seemed not just remote, but impossible. The spread of the first mills had been temporarily halted by riots. The same happened when manufacturers began to install power looms, throwing the handloom weavers out of work. The reaction to the power looms had the same root causes as the opposition to machine spinning, but the approach taken was very different. This was not to be an uncontrolled riot but a series of highly organised measures, based on the actions of the framework knitters of the Midlands.

The knitting frame for making hosiery had been in use for two centuries when the main knitting centres were the scene of a great outburst of frame smashing in 1811. Bands of knitters went out at night, heading for one selected workshop. Guards were posted, while their colleagues, armed with axes and hammers, set to work. It was all carried out with the precision of a military operation. The men used assumed names or numbers for the night, and their leader had a suitably military title – General Ludd. The followers became known as Luddites and the whole movement as Luddism. Today, Luddism is generally taken as a synonym for mindless opposition to progress, a simple antipathy to new machines and methods. This was not the case in 1811. Why should hosiers suddenly turn on the machines they had been using for generations? The problem was not with the machines themselves but the way in which they were being used.

Girls in the weaving shed at Holehouse Mill, Blackburn.

The argument was with the machine owners.

Theoretically, the hosiery industry was governed by a set of official rules, approved by parliament, concerning such matters as the proper employment of apprentices. With the movement of the industry from London to the Midlands, the rules became increasingly ignored. Apprentices were no longer considered as boys being trained in a skilled trade, but as cheap labour. The hosiers looked first to the due process of law. In 1778, they took their grievances to parliament where they told a sorry tale of exploitation. They dwelt especially on the apprentices: 'Some boys, who are Paupers, are put to work at the age of ten or eleven, but they make bad work of it – That the work effects them very much … the masters of these Boys make them work till Eleven or Twelve o'clock at night.' They complained about fraud and low pay and, in May 1788, they had to report that those who had come to parliament earlier that year had all been sacked on their return. Legal rights were demanded and promises made, but not kept. This was all too common at the time – others who went to law for redress fared worse. In 1812, Scottish weavers went to court to press for a minimum wage. They won the case, but the employers simply refused to pay. Faced with this flagrant disregard of the law, the weavers went on strike – and finally the law stepped in. The strike leaders were arrested, tried and imprisoned. The minimum wage remained unpaid.

It is against this background of legal methods tried and failed that Luddism must be seen. The Luddites decided on direct action, and the machine breaking began.

Some hundreds of country framework-knitters assembled in Nottingham Market Place, and expressed a determination of taking vengeance upon some

A contemporary view of Luddites.

of the hosiers, for reducing the established price for making stockings, at a time when every principal of humanity dictated their advancement.

The military were called in to break up the meeting, but that night another crowd gathered and sixty-three frames were smashed in Arnold, just outside Nottingham. A pattern was set. Frames were smashed at selected workshops by an increasingly well-organised secret army. Two magistrates came down from London and set up a secret committee with funds to pay informers, but none came forward. In 1812, frame breaking was made a capital offence, yet General Ludd's army still marched on the night streets, and even when more conventional trade unions were formed, frame breaking still broke out for time to time as a last resort.

Frame breaking was a direct attack on the employer, not a reaction to new machines and, whatever the morality of Luddism, there can be no denying it was more successful in achieving redress of wrongs than all the court cases and addresses to parliament had ever been. Employers had, in effect, been saying,

'We are sole arbiters of who shall work, how long you will work, where you will work and how much you will be paid for that work. Accept our terms or look for another job.' As there seldom were other jobs, this was no real choice. But the Luddites put forward a different agenda. 'If you try to increase your profits by wage cuts, sweated labour and the like, we will make sure you have no profits. We shall not stop work, for we know you can starve us out, but we will break the machines in which your capital is tied'. It was a crude answer, but it worked.

The story of the cotton industry was different from that of the hosiers. Here it really was about the arrival of new machines and the changes their arrival brought. For the employers, it made complete sense to install power looms that increased productivity, with one operative being able to take charge of three or more looms. At first, the spread of power looms had no significant effect, simply because there was so much yarn being produced, there was still work for all. It was different when trade slumped. The factory owners made sure that the power looms were kept busy at all costs. For the handloom weavers it was very different. Some owned their own looms, but found they had no work coming in, and it was even worse for those who rented their looms – for that rent was still demanded even when nothing was woven. Things came to a head in the trade depression of the 1820s. The weavers told their version of events in the popular ballads of the day:

Aw'm a poor cotton-wayver, as mony a one knows,
Aw've nowt t'eat in th'heawse, un' aw've worn eawt my cloas.
Yo'd hardly gie sixpence for a' aw've got on,
Meh clogs us' boath broken un' stockins aw've none.
Yo'd think it wor hard to be sent into th' world
To clem [starve] un' do best as yo' can.

Many ballads contrasted the lives of the weavers with those of their employers – a way of life caricatured in *The Hand-Loom Weaver's Lament:*

With the choicest of strong dainties your tables overspread
With good ale and string brandy, to make your faces red;
You call'd a set of visitors – it is you whole delight –
And you lay your heads together to make our faces white.

The chorus makes the new sense of militancy clear:

You tyrants of England, your race may be soon be run
You may be brought into account for what you've sorely done.

The conditions that gave rise to such bitterness can be found spelled out in the diary of William Varley, a weaver from Higham, a small village just to the north

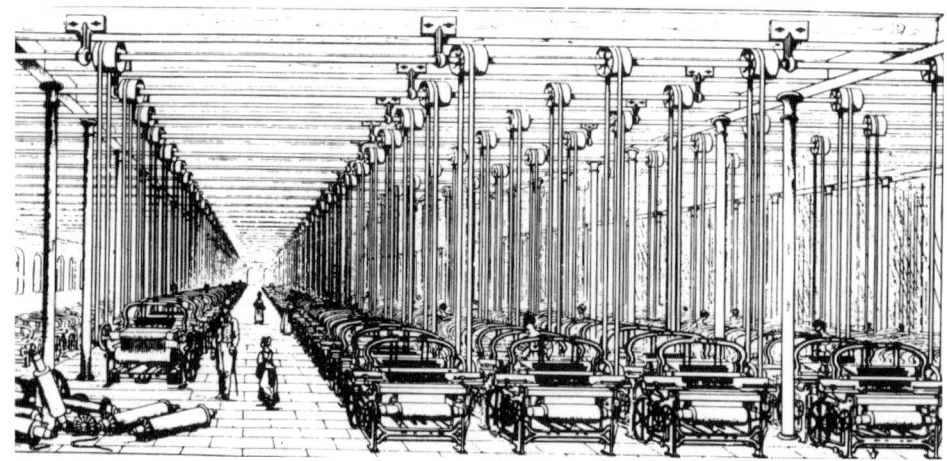

The vast array of power looms at Robinson's, Stockport: they could do the work of literally thousands of handloom workers.

of Burnley. He was luckier than some, for he had a country cottage with a garden, where he could grow his own vegetables and keep a pig and a few hens. Yet he found his standard of living falling drastically in the 1820s. From receiving 3s a piece for his weaving, the price had dropped to just 9d, though even that was better than the periods when there was no work at all. He was reduced to misery and the ignominy of having to ask for charity, and the hardest blow fell when his daughter died of tuberculosis. There is small wonder that his bitterness, and the mood is set in the first entry at the start of the decade. 'The year commences with very cold frosty weather … The poor weaver now very hard put to it, what with the rigour of the weather and the unrelenting hearts of our masters, whose avarice will not allow us above half wage.'

Not a new story perhaps; poverty, death and disease were far from unknown in the working population. What was new was the very plain fact that while men like Varley suffered, others were growing increasingly wealthy. This was not hidden poverty, unrecognised by those with authority, but when actions were taken to alleviate the poverty, the weavers found the actions were not altogether charitable. An unemployed weaver was given a job so that he could qualify for parish relief. He was ordered to deliver a heavy parcel to a village 10 miles away. On the way, he met another weaver with a suspiciously similar parcel. The two men opened the wrappings and discovered that what each of them was hauling across the countryside was a paving stone. They complained of the absurdity of this task to the parish overseer, who roundly abused them both for opening the parcels and refused them the promised relief. The faces of charity and cruelty could look surprisingly similar. Varley's diary goes on to record the darkening mood of the weavers in 1826.

March 11th Wm Hargreaves lowers wages 3d per cut and there is not half enough work so that the whole country is in uproar for the poor weaver cannot get bread

March 14th There is some disturbance at Blackburn this day; the poor people throw stones at the coach and break the windows.

In April there was a further cut in wages. In Bolton cuts were as much as 15 per cent and the more far-sighted of the manufacturers saw the dangers that lay ahead. They called a meeting at Bolton-le-Moors at which they deplored the action of those manufacturers who had taken 'an undue and unjustifiable advantage of the deplorable state of the depressed weavers'. Their resolutions had no force – the cuts remained. Perhaps if something had been done it would have had an effect on future events, but time and patience had both run out. Varley made a brief note on 24 April. 'This day the country rises in a great multitude and breaks the power looms of Accrington and Burnley and many other places.'

That short entry might suggest a wild, uncontrolled mob on the rampage, but there was little uncontrolled about the events of that year. The men had particular targets – the power looms and associated machinery. Their actions were roundly condemned by both the Radical and Conservative press. The latter put out the common argument in support of progress. It was unfortunate that the same progress seemed to be bringing poverty and misery to many, but such was the way of the world. 'There is, there can be, no other test of the intrinsic utility of a new machine, than whether it effects better or more cheaply, the purpose of that which has previously been in use. If it does, it ought to, and surely will, force its way.' The view is a logical one and one which probably manufacturers would accept today. The Radical press, while trumpeting the appalling conditions in the mills, describing mill girls reduced to prostitution and the starvation of the weavers, also argued that violence was not the answer – a view that would probably also be acceptable today. They also hinted darkly at the presence of paid spies, stirring unrest. That seems less convincing. Who would be paying them? And, given the conditions of the time, why would they be needed? Few watching the events of that April seriously believed that the advance of mechanisation could be permanently halted. Equally, given the enthusiastic support of authority for the mill owners, few doubted that the violence would be quickly brought under control, Was the machine breaking, then, no more than a futile attempt to turn back the clock? Not if you look at it from the weavers' point of view.

There was undoubtedly a great deal of resistance to the change from cottage to factory – and why not? Conditions in the mills were often wretched, but that was not the only problem. Weaving had always been a skilled, male occupation but now girls and young women could be employed to tend the power looms. The men lost work – and the women were on low wages. Everywhere the machines seemed to be benefitting the few to the detriment of the many. Even if that was true, was machine breaking the answer? There was no shortage of commentators to point out it was at best a short-term solution to the problem. But it was an immediate problem that faced the weavers. If they smashed the looms but left the spinning machinery untouched, then they would have to send the yarn out for weaving. This is precisely what thousands did.

'No damage has been done to the spinning parts of any of the factories,' reported the *Bolton Chronicle*, 'nor even to the windows, the sole object of attack being the power looms.' The report concluded, 'no disposition to commit outrage upon any other property has been manifested'. If this was rioting, it was remarkably restrained rioting.

On Wednesday, about 100 of the rioters assembled between Haslington and Rawtenstall, about ½ past 8 o'clock in the morning. They gave a shout at the north end of the hill leading to Rawtenstall, and immediately descended into the village, and commenced an active attack upon Mr. Whitehead's factory, which was strongly barricaded on the inside, and occasioned them a delay of half an hour, before they could effect an entrance. On entering they immediately

The magnificent steam engine at Coat's Paisley. The ropes wound round the flywheel in the centre of the picture carried the drive to all parts of the mill.

commenced the work of destruction, and in the short space of half an hour they completely destroyed ninety looms – all that the building contained. They did not attempt to injure any other machinery besides the looms. On coming out they gave three cheers, and proceeded to Mr. Thomas King's of Long Holme.

They seemed to be a remarkably controlled group. After visiting the Rawsthorne factory at Edenfield, someone remembered the dressing frame, another machine involved in cloth making, so back they went, 'but on Mr. Rawsthorne assuring them there was nothing but spinning machinery, they retired without doing any further damage'.

The end was inevitable. The military were called in, lives were lost, injuries sustained. The bitterness and despair were on plain view. 'Our informant saw a poor fellow lie down between the feet of the soldiers' horses: he said they might trample on him if they liked, he was starving to death, but he would persist in breaking the looms.' The violence was soon over but had had its effect. The condition of the weavers became a matter of public concern, as reports began to appear describing conditions in the cotton towns that had led to the riots. This report in the *Bolton Chronicle* of 1826 describes conditions in Colne:

This place, like others mentioned, exhibits one scene of equal distress. Work for weavers is so scarce, that the night before the taking-in day, the weavers come at eight or nine o'clock in the evening, and remain starving with hunger and cold, until eleven or twelve o'clock the next day. The wages which they are receiving are so mean, that you would wonder how the poor creatures could exist. There are employers here who give one shilling and threepence per cut of thirty yards – other one shilling and three-halfpence and some others only one shilling. Report says that the present overseer of this town, made an attempt to live for a week upon one shilling and sixpence (the parish allowance to each person) and it was consumed by Thursday night. His diet was porridge and treacle, and he declared that this allowance could not keep any human being alive for any length of time.

Government relief was sent. Varley received 9lbs of meat on 23 May, and the next day brought even better news. 'This day I got work of Mr. Corless, so now I hope through the mercy of God I will be able to maintain life a little longer.'

The relief for individual workers was limited, but the riots did at least focus attention on the textile districts. Parliament set up a Select Committee to study the problem, and they came up with a solution – export it, send the poor weavers off to the colonies. That idea was quickly dropped when it was worked out that the poor could not afford the fares, so the government would have to pay. The investigations also brought more unsavoury facts to light, in particular that things were not only bad in Lancashire, there was similar poverty in the cotton

Cotton operatives clashing with the military in Preston, 1842: two workers died.

industry round Glasgow and Paisley, where there were estimated to be some 11,000 handlooms and the average weekly earnings were only 5s 6d. And it was not just the 'outdated' handloom weavers who were suffering, Things were just as bad for those using the new, efficient machines. Progress was not making their lives better. The figures below relate to a man working with power looms, described by his employer as 'a first-class workman. Provisions for a year came to just over £23 of which these were the main items:

Meat £6 10s 0d
Potatoes £3 18s 0d
Milk £4 11s 0d
Bread £2 1s 2d
Soap 17s 4d
Starch 12s 0d
Coal £1 18s 6d
Oil 16s 6d

On top of that there were items such as rent at £4-10s-0d, medicine, clothes and shoes for the family. The grand total for a year's expenses came to £32 12s. To set against this was his income of £26 a year and his wife's pay, a mere £3 18s, making a total of £29 18s. It needs no Mr Micawber to point out that the net result of that financial equation is misery. The same story was repeated in every district where cotton was spun and woven. Things were bad and getting worse. In 1815 a 'second-rate' weaver could earn £45 a year. In 1826 a first-rate man had to work sixteen hours a day to make half that amount.

Where was the prosperity of which everyone talked? Where was the progress? It seemed that progress in the industry did not translate into prosperity for the workers. The Select Committees sat, pondered and did nothing. The Relief Committees passed out their meagre rations to the poor. The textile workers were starting to adopt a new attitude towards their employers, since it seemed increasingly clear that they were not going to give anything away as a right. The two sides of industry were moving apart, taking up increasingly hostile positions. A working class was emerging, aware of its own identity and increasingly convinced that the only way forward was through unity and sheer weight of numbers. The awareness did not arrive like a flash of light on the road to Damascus, but rather as a gradual dawning. Even though the message was becoming increasingly clear, it was no easy matter to put it into practice.

The idea of workmen co-operating to help each other through the bad times was not new in the 1820s. Thirty years before, the spinners had banded together in Benefit Societies. They had got together to petition parliament about low wages; they had given evidence before Select Committees, produced documents to support their cause and sought the support of men of affairs. The parliamentarians listened, recorded the arguments and did nothing. The petitions did, however, make them aware of a new tendency for workers to get together to make demands. They responded with the Combination Acts, making it illegal for workers to get together to pursue pay claims. The Acts also forbade employers from getting together to fix wage rates, but while several workers' organisations were prosecuted for banding together, no employers were ever charged. The Combination Acts were short lived, largely because they proved ineffective, and actually encouraged the growth of trade unionism. For the workers, the Acts reinforced the view that the government was unlikely ever to take their side, but was far more likely to act against them. The only time authority was seen to act was when troops were sent in to support mill owners. Employees realised they could only rely on their own resources.

In 1829, there was a series of strikes against a reduction in wages. The leader of the Lancashire spinners was John Doherty, a 30-year-old Irishman. That year, he organised the biggest of the strikes in Hyde, near Manchester. It lasted six months and was marked by a new bitterness and by a new sense of defiance and purpose, spelt out in a handbill: 'You are mistaken if you suppose that 14 weeks' starvation can show us the propriety or reasonableness of a measure which we could not see before and the mischievousness of which we can demonstrate.' The new militancy was apparent in the correspondence between the two working class leaders. The Chartist Francis Place wrote to Doherty, criticising a pamphlet:

A large portion of the printed address is worse than useless, in as much as it is an appeal to the humanity of the master, against their interest. The reasons given for the appeal are futile. The manufacturer looks only to his immediate profit, and cares little or nothing for what may be the state of trade hereafter, or what may be the profits of his successor ... Depend upon it the working people never will, as they never have, obtain anything by such appeals. The struggle is a struggle of strength and 'the weakest must go to the wall'. Whatever the people either gain or even retain, is gained or retained, and must always be gained or retained by power.

It seemed a bleak prospect of perpetual struggle, but for all the talk of militancy and holding out forever, the strike failed. Six months of near starvation broke the strikers, but one new lesson had been learned. A single district could not hope to win a strike – combination and concerted effort throughout the industry was needed. If wages were cut in Glasgow, for instance, then the Scots would receive help from Manchester, and vice versa. In December 1829, the first steps were taken towards a national union when members from England, Scotland and Ireland met to discuss the problem. They agreed to form the National Union with Doherty as secretary and passed a number of resolutions, demanding for example a reduction in the working hours for those aged under 21. Authority listened to their deliberations – there was nothing secret about their meetings – and reacted in panic, writing to Robert Peel, the Home Secretary, in May 1830:

The combination of workmen, long acknowledged a great evil, and one most difficult to counteract, has recently assumed so formidable and systematic a shape in this district we feel it is our duty to lay before you some of the most alarming features ... a weekly levy or rent of one penny per head on each operative is cheerfully paid. This produces a large sum, and is a powerful engine, and principally to support those who have turned out against their employers, agreeable to the orders of the committee, at the rate of ten shillings per week for each person. The plan of a general turnout having been found to be impolitic, they have employed it in detail, against particular individuals or districts, who, attacked thus singly, are frequently compelled to submit to their terms, rather than to the ruin that would ensue to many by allowing their machinery (in which their whole capital is invested) to stand idle.

The authorities over-reacted, and certainly overestimated the abilities of the cotton workers to form such a united group. Communications between the two major centres, Manchester and Glasgow, were poor, and even in its heyday the union did not include the women and children who made up the bulk of the workforce. By no means all the workers were convinced of the value of unions – and nor were all union official scrupulous in their methods of recruitment. The darker side of the movement was hinted at in 1838, when five Glasgow spinners were put on trial, reported here in the *Annual Register*:

The cotton spinners of Glasgow have long been noted for the violent and arbitrary proceedings of their confederacy. Early in January, five individuals connected with this body, were indicted at Edinburgh on counts charging murder, attempts at arson, and conspiracy, besides other grave offences of a similar nature.

It all sounds most sinister, but after the trial had already begun, the prosecution introduced three new counts: conspiracy; illegal combination; and writing threatening letters. Ninety-one witnesses were called by the prosecution, who described 'some curious though revolting details of the practices and formidable reputation of the cotton spinners union of Glasgow'. At the end of the day, they were acquitted of all the serious charges, and only found guilty – and that by a majority verdict – of the three lesser charges hastily introduced during the trial. One can only assume that the prosecution had belatedly realised that their original case was unlikely to stand up to questioning. Nevertheless, even though the men had already served five months in gaol, they still received the harsh sentence of seven years transportation each. The lesson was obvious: the authorities were determined to break the new unions.

An anti-union cartoon of the 1830s. This was very much the accepted establishment view of unions.

The great trial of strength between owners and workers came in 1853. 'Preston has become somewhat celebrated', wrote the town's historian, G. Hardwick, 'as the principal "battle field", where capital and labour engaged in the cotton manufacture fight in defence of what each deems its respective rights and privileges.'

The lead up to the big strike was long. Wages had been cut in the recession of 1847, but promises had been made to return to the old levels when trade improved. Trade did improve, but against a background of rising food prices. The operatives put in their demand for a 10 per cent pay rise, and here the story becomes confused. Some mills gave the full rise, some argued about the amount, and some gave nothing at all. Then a single, clear issue emerged; the workers demanded 10 per cent for all. The masters dug in their heels, declaring that no one would be allowed to dictate terms to them and, in October 1853, closed down the mills. The lock-out had begun. From being a negotiable argument over pay, it had become that far less tractable problem – an argument about principles, as Henry Ashworth who wrote a contemporary account makes clear:

The real question at issue in the Preston strike was not one of wages, but of property: not whether the operatives would have more or less money in exchange for their labour, but whether the masters should have the power of saying whom they would employ and on what terms; whether they would be masters within their own just province, viz. within the factories they had built, and among the men who received their money. Their cause was not of capital against labour, but that of property against communism.

The strike was marked by bitterness. From the start, the masters refused to accept any form of arbitration, arguing that no outsiders had the right to meddle in their affairs. More than 20,000 workers were laid off, but the mills did not stay idle for long. The 'knobsticks' came to Preston. They were mostly the poor of the agricultural areas of Ireland, who were recruited for the work, together with the few Lancastrians prepared to work on. Their presence did nothing to calm the situation, and nor did the action of the authorities who began arresting union leaders on conspiracy charges. The workers sent a deputation to the Home Secretary, Palmerston, who replied with a lecture on laissez-faire economics – wages are determined by supply and demand, and morality does not come into it – and sent them packing.

The mood of defiance can be heard in many of the broad sheet ballads published at the time:

You may see, of a truth, that the people are not dead,
Though 'tis said they died long ago;
We've risen from our sleep, a holiday to keep,
Determined to work under prices no more.

So we've thrown away reed, hook, and comb,
And hung up the shuttle on the loom;
And we'll never be content, till we get the ten per cent
In spite of the 'let well alone'.

The ballads did more than cheer up the strikers. One of the most famous 'the Cotton Lords of Preston' was hawked around the district to raise funds:

So men and women all of you
Come ad buy a song or two
And assist us to subdue
The Cotton Lords of Preston.
We'll conquer them and no mistake,
Whatever laws they seem to make,
And then we'll get the ten per cent
Oh then we'll dance and sing with glee
And thank you all right heartily,
When we gain the victory
And beat the Lords of Preston.

The strike lasted for thirty-six weeks, right through the winter, and for all the brave words of the ballads, there was little hope that the operatives could win. Funds became short, and the effect of the knobsticks was felt very strongly. There was a slow drift back to work, and the union leaders had no choice other than to capitulate. The owners celebrated their victory.

It is now surely established, on the basis of experience, as well as reasoning, that, no more than any other commodity, can labour obtain a higher price than the purchaser is willing to give for it. This is a law in which, whatever the consequences may be, we have to acquiesce, just as much as we should to the law of gravitation ... But happily, it might be demonstrated, that all the results of this economical law, are in perfect harmony with justice and benevolence.

Others, even when they were sympathetic to the masters, saw matters rather more soberly and doubted that the end result was 'perfect harmony'.

The victory obtained by the employers merely demonstrated that which everyone previously knew, viz., the strongest party in the end would win. But this is not sufficient to set at rest the mighty question, which yearly throbs with increased vitality beneath the surging mass of mercantile contention. No one really wins in these struggles. They are essentially productive of loss to all, except in so far as they inculcate lessons of wisdom. It is, therefore, the duty and

A trade union banner.

interest of all that the differences which must occur occasionally between the buyers and sellers of labour, as well as of any other commodity, should be settled in a commercial, not in a military spirit.

This was all very well in theory, but in practice, as seen at Preston, there was no attempt to settle by reasonable discussion between equals. Even the latter writer regarded the workers and their labour as 'commodities' – and you don't have to consult commodities. Was that really so very different from the attitude of the planter to his slaves? And if the rising of the whites was not to be, what hope was there for the 'insurrection of the blacks'?

Chapter 13

THE INSURRECTION OF THE BLACKS

If the white cotton operatives of Lancashire were faced with apparently insurmountable difficulties in obtaining a decent standard of living and a dignified way of life, then the obstacles faced by the black rose even higher. They had no more status than a cow in a meadow or a chair in the master's house. They were simply objects to be bought and sold. They could no more appeal to a court of law for justice than a chair could sue an overweight man who broke its leg. The British operative found the law to be generally unhelpful, but at least it did exist and the possibility of using it was there. Not for the slave, for whom the law was made by white men for white men. If the slave was injured, the master could sue for compensation, but if he won the claim the compensation was his alone. For the mill worker, the possibility of change did exist, at least in theory. He or she could leave the mill and perhaps make a fortune as the preachers of the Victorian doctrine of self-help never tired of pointing out. The mill worker could, as a few did, move up the social ladder and leave the past behind. The slave carried the mark of his subjection with him, wherever he went, in the colour of his skin. What hope was there for any change in his life? Could he do anything to change his position?

The only possibility appeared to be revolt, a rising up against the plantation owners. Those who wonder why the slaves did not, in fact, rebel have not really thought about it from the slave's perspective. Start to add up the obstacles in the way of rebellion and it is remarkable that there were any uprisings at all. Compare the positions of slave and mill worker. The mill workers were able to organise with others in different districts, even if communication was difficult. The slave was lucky if he was allowed to talk to a member of his own family in the next plantation. The British worker could rely on support from many of his neighbours but the slaves were surrounded by a hostile population, who regarded them as something less than human. Yet, in spite of the difficulties, there were revolts and their effect was out of all proportion to the scale.

The largest of the slave revolts is, in fact, the least well known. In 1811 some 500 Louisiana slaves marched on New Orleans. They put up a brave show, striding along with banners flying. But hoes and axes were no match for swords and guns; defeat came with swift inevitability. It caused little stir in the South. Louisiana had only been purchased from France eight years earlier and was still regarded as a foreign land. A few years later, it might have seemed more of a threat. There were two other abortive risings, led by Gabriel Prosser and Denmark Vesey, but there was one revolt that attracted huge notoriety and prevented many a Southerner from enjoying a sound night's sleep, even after it had ended. It was the rebellion of 1831, led by the Virginian slave, Nat Turner.

The revolt might seem to be quite a small-scale affair; there were never more than 100 slaves involved, compared with the 1,000 who got together in the Preston strike. But it was not the numbers involved that caused panic, but the fact that sixty whites were killed – though 100 blacks died in the aftermath. Turner himself remained free for two months after the rising had been quelled. He was captured on 30 October 1811, ands two weeks later he had been hanged. He left behind a personal statement, giving some details of his own life and a chilling account of the rebellion itself.

Nat Turner was born in 1800. He gave no details of his parents, though both his mother and grandmother feature at the start of the narrative. His father is

Nat Turner discovered in hiding after the collapse of the slave revolt he had led.

not named, though we learn he ran away – but whether he escaped to the North or was captured is not known. Only one man is named in this part of the story, his owner, Benjamin Turner. At the very start of his life, Nat was denied a part of his identity. The owner chose the name 'Nat' and added 'Turner' as a mark of ownership. His name was not something of which he could be proud but a badge of inferiority. Not all slaves accepted the situation. When Mammy Maria was asked why she called another slave Henry, 'Mister Ferguson', she replied angrily, 'Do you think 'cause we are black that we cyarn't have no names?'

Religion played an important part in providing the cohesion in the slave community that the law discouraged. It offered an authority higher than that of the owner, a court of appeal more potent than any human court. The white man was always able to quote Scripture to justify slavery; the slave used that same Scripture to attack it. In the earlier Vesey rebellion, one of the slaves, Rolla, gave evidence to the court:

> At this meeting Vesey said we were to take the Guard-House and Magazine to get arms: that we ought to rise up and fight against the whites for our liberty; he was the first to rise up and speak, and to read from the Bible, how the children of Israel were delivered out of Egypt from bondage.

Nat Turner went further than this, claiming a Messianic mission. Reading his own words, one is struck by the obvious depth of his convictions – and by the mixture of older African beliefs with his Christianity. He described how, as a child of 3 or 4 years of age, he had begun talking about events that happened before he was born:

> Other who were called on, were greatly astonished, knowing that these things had happened, and caused them to say, in my hearing, I surely would be a Prophet, as the Lord had shown me things that had happened before my birth. And my mother and grandmother strengthened me in this my first impression, saying, in my presence, I was intended for some great purpose, which they had always thought from certain marks on my head and breast.

Later he was to claim divine inspiration for the uprising, receiving his orders directly from 'the Spirit'. His questioners came to Turner his condemned cell and asked him if he did now feel that he had been mistaken. He simply replied; 'Was not Christ crucified?' There is no doubt that he believed himself to be divinely inspired and so there is no hint of remorse in his account of a typical fatal raid on a plantation:

> I took my station in the rear, and, as it was my object to carry terror and devastation wherever we went, I placed fifteen or twenty of the best armed and

most to be relied on in front, who generally approached the house as fast as the horses could run. This was for two purposes – to prevent their escape, and strike terror in the inhabitants; on this account I never got to the houses, after leaving Mrs. Whitehead's, until the murders were committed, except in one case. I sometimes got in sight in time to see the work of death completed; viewed the mangled bodies as they lay, in silent satisfaction, and immediately started in quest of other victims. Having murdered Mrs. Waller and ten children, we started for Mr. Wm. Williams', - having killed him, and two little boys that were there; while engaged on this, Mrs. Williams fled and got some distance from the house, but she was pursued, overtaken, and compelled to get up behind one of the company, who brought her back, and, after showing her the mangled body of her lifeless husband, she was told to get down and lay by his side, where she was shot dead.

This mixture of religious fervour and ruthlessness shocked the South. Every planter family saw themselves surrounded by Nat Turners. There were inevitable repercussions, and many slaves came to resent the trouble stirred up by Nat Turner. 'We poor colored people could not sleep at nights for the guns and swords being stuck in at our windows and doors to know who was here and what was their business.'

Slave revolts were few but that does not diminish their significance. Although the possibility of revolt was, to say the least, limited, individuals had their own ways of making protests.

A constant complaint made against the blacks was that they were idle, dishonest and stupid. If idleness is seen as a refusal to do more work than is absolutely necessary in a situation where there is no reward for work, and if dishonesty is seen as taking some sort of reward for that work when the master refuses to do so, then perhaps the case for stupidity has already been rebutted. It has often been pointed out that that the work ethic so strongly recommended by the masters was not one that they chose to apply to their own lives. The slave, seeing his master enjoying a life of ease, might be forgiven for feeling that he would settle for a little less virtue for himself and a greater share of sinful comfort.

We tend to regard all forms of stealing as being fundamentally immoral, but visitors noted that the slaves had their own system of morality:

It is told me as a singular fact, that everywhere on the plantations, the agrarian notion has become a fixed point of the negro system of ethics: that the results of labor belong to the laborer, and on this ground, even the religious feel justified in using 'Massa's property' for their own temporal benefit. This they term, 'taking', and is never admitted to be a reproach to a man among them that he is charged with it, though 'stealing', or taking from another man other than their master, and particularly one another, is so.

Stealing from the whites was justified over and over again in the testimonies of former slaves after the Civil War. The whites, they said, were the biggest thieves of all, for they had stolen the people from their homes in Africa. So, they 'took' and made this a form of protest. But the most common protest was the most direct of all – they ran away.

The runaway slaves fall into two categories: those who hoped to make it all the way to the North and freedom, and those who left for reasons of their own, and often returned of their own free will.

The Underground Railroad was the most famous example of an organisation set up to get slaves out of the South. They were passed from 'station' to 'station', their every move controlled by agents and many, such as the magnificently resourceful Harriet Tubman, became folk heroes. Though many travelled the railroad it could only have been a tiny minority of the whole slave population.

Runaway Slave—Runaway from the subscriber No 221 Magazine three weeks ago the negro woman slave named SOPHIA, aged 27 years, about 5 ft high, marked with the small pox, crooked feet; big lips, wants some teeth before, was dressed when she started with a blue spotted domestic frock, she is well known in suburb St Mary as a washer by the day, and is supposed to have been harboured in said suburb.

Ten dollars reward will be given for the apprehension and delivery to the subscriber of said slave or for lodging the same in the jail of New Orleans.
m 31–3t   P. SHIELD.

FIFTEEN DOLLARS REWARD.
Run away from the subscriber, Esplanade corner of Rampart street, on the 28th inst, the negro girl SARAH, 19 years, 5 fhet 2 inches high, she has a sulky look when spoken to; when she left was dressed in a light purple and spotted calico frock. She has a large scar between her shoulders. The above reward will be paid to whomsoever will will apprehend said slave and lodge her in the parish jail of New Orleans.
mar 31   JAMES FINLAY.

NOTICE—Detained in the jail of the parish of Jefferson, a negro boy who calls himself John, is about 12 years old, and says he belongs to Mr Williams. Also a mulatto boy called, Anfield, 14 years old who says he belongs to Mr Bouligny. The owners are requested to claim them in conformity with the law.   J. CHARBONNET,
m 29   Sheriff of the parish of Jefferson.

Newspaper adverts for runaway slaves.

It was, however, important as a safety valve, relieving some of the worst pressures of the system. The bitterest fact for many was not just that they were slaves, but their children too would be slaves, and their children's children for generations to come. Every slave that escaped opened up a chink in that gloomy future, offering a glimpse of freedom.

For every slave who found a new life, there were thousands who made temporary escapes. Some were gone for months, some for a matter of days. Those who fled in the hope of getting clear away face appalling difficulties as a former slave described:

> No man who has never been placed in such a situation can comprehend the thousand obstacles thrown in the way of the flying slave. Every man's hand is raised against him – the hounds are ready to follow on his track, and the nature of the country is such as renders it impossible to pass through it with any safety.

The odds against the slaves were indeed formidable. Many owners kept packs of 'nigger dogs' to hunt down escaping slaves, and too often the hunt ended with the dogs savaging their prey. Most slave catching fell to the slave patrols, made up of local poor whites. As a group, they were universally detested in the slave quarters, and with every justification, for they were noted for their brutality. That became even more pronounced as abolitionists became more vocal. The runaway, who had previously been thought of as a special kind of thief of his master's property – stealing himself as it were – now became a dangerous part of a hated political movement. Repression increased as the abolitionist movement grew.

This, however, was only one, if the most dramatic, part of the runaway story. Every plantation seems to have had its share of runaways – and not all of them even attempting the long trek to the north. For many it was a form of protest, the only one open to them. The diary of John Nevitt, who owned a plantation in Mississippi, is dotted with accounts of slaves running away, being captured or returning on their own. It is worth looking at in some detail because it illustrates just what a complex matter this was for both owner and slave.

Entries are generally brief, with no indication of a slave's motives. These entries appeared in the early part of 1826:

February 2	Peter run away in the morning
April 24	Pete who ran away 2nd Feby last was Brought home by Mr. Dreggs paid him his fee, and whipped and ironed Peter.

This is the traditional idea of the runaway, brought back by the slave patrol, punished and placed in irons. The following year, there were some very different entries:

February 28	Bill ran away
March 3	Bill came home last night.
April 21	Maria run away.
April 28	Bill & Jerry catched Maria at Campbells.
April 29	whipped Maria and put iron on her leg.

Here we have two very different phenomena. Bill awards himself a short holiday: Maria disappears to another plantation. To meet a lover, to plan an escape? We are not told, but after only a week she is caught, not by the patrols but by fellow slaves, whipped and put in irons. Yet, in contrast to the harsh treatment handed out to Maria, we find this reference to John, whose job was carting:

| July 20 | John who ran-away on the 1st Inst came home forgave him and set him to work had his team got in readiness for tomorrow |

In spite of the lenient treatment, John was off again two weeks later. Harsh treatment was no more effective at keeping the runaway at home. There was no holding Maria:

21 August	Maria who ran away on the 12th was brought home by Rubin gave her a light whiping and set her to making cotton bags
28 September	Maria runaway
1 October	Rubin runaway
2 October	Rubin came home in the morning sent him out for Maria he returned in the evening with her forgave Rubin his falt and gave Maria severe whiping.
9 November	Maria was brought home by Jerry on the night of the 6th had her whiped severely an Ironed with a shackle on each leg connected with a chain.

So the story goes on, with both Rubin and Maria regularly running away and regularly being caught. Only once, in March 1828, was it recorded that Maria came home voluntarily, and on that occasion she escaped punishment. There is no hint of the depths of despair that sent her off time after time, in spite of the punishments every time she was caught. What the bare account does, however, is point out the old dilemma facing the planter. If the work had been done by hired labourers, he could have let them go and hired others in their place. But Rubin and Maria were not employees, they were a capital investment, just like the machinery of a mill. A runaway slave represented a loss of capital – but a shackled slave was not going to be very productive either. In the case of Maria, the final answer came on 26 December 1828 when she was sent to Natchez Gaol.

The mule driver: an important man in plantation life.

Whatever that may have meant for Maria, it was an admission of defeat by Nevitt. His property was locked away, useless.

The problem of Maria was solved, though that of Rubin remained. His name keeps cropping up right through to 1830. 'Went to the swamp to hunt runaways … several men came from town to hunt runaways the overseer with them. Took little Sal who was in company with Rubin, Sandy.'

And on 11 January this entry appears. 'Rubin came home having been runaway for two or three months – *did not punish him.*' The emphasis is Nevitt's, clearly impressed by his own generosity, though there is no indication why he was generous to Rubin, while beating and eventually gaoling Maria. If running away was one means of opposing the master, there were other more violent measures. On 13 January 1827, the cotton gin was destroyed in a fire, and arson was suspected. But as the culprits were never found, there is nothing to prove it was the work of slaves. Indeed, there is a real possibility it was not, since one of the slaves, Gusty, died in the blaze. Maintaining a balance between leniency and punishment was always difficult. Sometimes the problem was taken out of the owner's hands:

> 27th July 1830 rode to Natchez found that Kate was put in Jail on suspicion of
> having stolen meat found she had received it of Bill rode out with Mr Armstrong
> (the constable) and apprehended Bill who after a little flogging confessed that

he and Mrs Campbells Sandy had broken into Mr Lyles warehouse and taken from it Bakin & Liquor Mr Armstrong also apprehended Sandy took them to both to Jail.

Bill was found guilty and sentenced to thirty-nine lashes, and now Nevitt had to decide what to do with him. He left him in gaol while he considered the matter, and forgave Kate her part in the business, since she had testified against Bill. Sandy is presumably the one who was earlier brought back as a runaway. It was the old problem – which was better, the carrot or the stick? And the same conflict – the owner denied the slave his humanity – the slave asserted it. The image of the 'good, benevolent planter' became a part of Southern mythology. During my travels through the South, I met several descendants of planters who, without exception, were prepared to admit that there had been some bad owners, but their ancestors had all had excellent relationships with their slaves. But the system was also a breeding ground for racial hatred to spread throughout the white population and could bring poor whites into conflict with the planters. One example arose when a ferryman got into an argument with one slave, America, and proceeded to take it out on other slaves. The plantation overseer, Peter Jaillet, told the subsequent story to Major J. Crawford, the owner:

It seems that Mr Moran either dissatisfied at the escape of America or determined to wreke his vengeance on some of my slaves – took occasion on last Monday to whip with a Wagon Whip & beat with a hand spoke another of my men – named Tom – whom I had sent to my farm – upon being informed of this circumstance on my return from the farm whither I had gone early on the same afternoon I enquired of Mr Moran the cause of such correction – he replied that the slave had 'saucied' him. That he had beaten him, that he intended to beat him again upon his return & would beat up him or any other of mine when it pleased him – I remonstrated with him as to the mode of correction, stating that if one of my slaves ever gave me offence he should be punished severely for it – but I wished him corrected properly and lawfully – not with bludgeon and hand spike, which might disable or injure him as to render him unfit to do his duty – he thereupon reiterated the threat to beat him upon his return & added (with no small insult to myself for protecting, as he termed it, such saucy negroes) that he would beat him with a hand spike or anything else he could lay his hands upon. In consequence of this threat and such treatment Tom went by Holts ferry, where the principal part of my crossing has since been done at an expense which I cannot well defray.

The antipathy of slaves towards their masters showed in many different ways. They were often accused of slacking, but cotton picking is back-breaking work and the demands were extreme. 'Commenced picking cotton on the 1st day of

The busiest, and the hottest, time of the year – cotton picking.

August 1860. Dry weather. Thermometer at 106.' Even in these temperatures they were each able to pick 1,000lbs of cotton a week per person.

There was some relief from the life of drudgery, and one particular solace was in religion, with its promise of a better life to come: they took comfort in the human fate that most of us dread:

> We're a marching to the grave
> We're a marching to the grave, my Lord
> We're a marching to the grave,
> To lay this body down.

It says a good deal about the human life that this was considered a hopeful and very popular hymn.

They also created their own society, which often seemed comical to outsiders. Underneath the mockery of the following account from the pen of John A. Quitman we can still get a picture of a group establishing its own identity and values:

> At a wedding I witnessed here last Saturday evening, where some 150 negroes were assembled, many being invited guests. I heard a number of them addressed

as governers, generals, judges, and doctors (the titles of their masters) and a spruce, tight-set darkey, who waits on me in town, was called 'major Quitman'. The colored ladies' are invariably Miss Joneses, Miss Smiths, or some such title. They are exceedingly pompous and ceremonious, gloved and highly perfumed. The 'gentlemen' sport canes, ruffles, and jewelry, wear boots and spurs, affect crepe on their hats, and carry huge cigars. The belles wear gaudy colours, 'tote' their fans with the air of Spanish senoritas, and never stir out, though black as the ace of spades, without their parasols. In short these 'niggers', as you call them, are the happiest people I have ever seen.

It was common for the masters to give holidays on 4 July and Christmas Day. The latter was a special treat and considered very important by the masters. 'I have endeavoured to make my Negroes joyous and happy – & I am glad to see them enjoying themselves with such a hearty good will', wrote Everard Baker, adding in his diary three days later:

I have recommenced work today – I called all hands up last night, told them the work we had before us compelled our holidays to close, & made a few remarks to them as to their duties the following year. They seemed thankful for the favor I had extended to them & eager to commence work I gave them from Thursday last night ... I did all I could to make their holidays pleasant to them & they seem to appreciate my endeavours.

Later in the year he had another friendly occasion to report. 'Wm & Rachel married last night – I performed the ceremony. Gave them a nice supper, had several of the darkies from around here – I played the violin & they danced awhile, everything passed off pleasantly.'

He was obviously one of the genuinely better masters, but the same diaries also tell a different story. There are accounts of violence, of a slave attempting to stab an overseer, and the familiar comments about using the lash. Somehow the happy days and holidays appear less frequently than the beatings. For some, faced with the bleakness of a life of slavery, there was only one solution. Too many diaries had entries such as this: 'Cherry found hanged by the neck.'

Chapter 14

SCAPEGOATS

The slaves of the cotton plantations were in no position to change their status. They could and did temper their conditions in a variety of ways, but it was totally outside their powers to alter the core of the system – slavery itself. For that to happen would require either a dramatic change in attitude by slave owners, a highly unlikely event, or a sustained campaign by white abolitionists and the legislature. The plight of mill workers was, on the face of it, less desperate. They were free members of society who, if they found factory work wholly abhorrent, held the remedy in their own hands – they could leave and look for work elsewhere. As they did not do so, the general assumption was that they were reasonably content; the alternative explanation was that they were well aware

there was not necessarily other work to be had, or if there was it offered even worse conditions. The proponents of the factory system regularly put forward the view that life for the factory worker was far better than that enjoyed by others of their class in the country.

To decide whether life was better for the textile workers under the new system or under the old is, to say the least, a complex matter. For a start, the industry was subject to fluctuations of slump and bust and that makes comparison difficult. No one, however, would seriously dispute that the life of the

A former slave cabin, photographed in Georgia c.1900.

average worker was improved in the long term by the process of mechanisation. But during the period of transition, things did not look quite that simple, and if one looks at the question in terms of something other than just earnings, then the period of change caused long-term damage. A gulf developed between masters and workers, not just between prosperity and poverty, but a gulf of mutual incomprehension. There seemed to be a total inability on the part of many masters to see the system from the operatives' point of view. Andrew Ure, the chief advocate for the industrial world, looked at the new age saw it as already in a state close to perfection:

> The constant aim and effect of scientific improvement in manufacturers are philanthropic, as they tend to relieve the workmen either from niceties of adjustment which exhaust his mind and fatigue his eyes, or from painful repetition of effort which distort or wear out his frame. At every step of each manufacturing process described in this volume, the humanity of science will be manifest.

Yet, for some reason, the workforce seemed not to appreciate that they were being offered paradise:

> Even at the present day, when the system is perfectly organised, and its labour lightened to the utmost, it is found almost impossible to convert persons past the age of puberty – whether drawn from rural or from handicraft occupations, into useful factory hands. After struggling for a while to conquer their listless or restive habits, they either renounce the employment, or are dismissed by the overlookers on account of inattention.

Modern readers of Ure will feel that his first description of the beneficial effects of mechanisation is little more than an account of work from which all interest has been removed, but his failure to understand the reasoning of others went beyond this. He later described one of 'our enlightened manufacturers' of Stockport who had introduced this beneficial new system. He shared Ure's enthusiasm for labour-saving machinery and was bewildered by the lack of equal enthusiasm among his employees. It seems not to have occurred to Ure that one of the great benefits of the system – that 'he would save £50 a week in wages, in consequence of dispensing with nearly forty male spinners' – might be less enthusiastically received by the men thrown out of work.

Ure, if he had thought of such matters, could not doubt that this was the price that had to be paid for progress, for increased productivity and expanding trade that would ultimately benefit all. The workers could, and did, argue that they saw no reason why they should be the only ones to pay that price. The crucial argument is not about whether the system was good or bad, but about the way in which it was introduced. The factory system was imposed on the many by the few, working conditions were laid down by the few on the many – the changes were welcomed by the few, not the many. However convincingly you can argue that the changes were for the best, there is no denying they were unwelcome to the vast majority and pushed through with little or no regard for the effect on their lives. One imposed – the other accepted, reluctantly. The result was a general hostility to change.

The resentment felt towards the factory system was not without cause. The hours were long, pay low and conditions bad. Increasingly, much of the labour went to women and children. The argument that things had always been bad, or that they were no better in other places, was largely irrelevant. In an industry supposedly heading full speed towards Utopia, the workers were entitled to ask – why aren't things getting better? The apologists came up with two answers. Firstly, things really aren't as bad as they seem, don't be deceived by appearances. 'Mr. Wolstenhome, surgeon at Holton, says that "The health of the factory people is much better than their pallid appearance would indicate".' Where, however, things really were very bad indeed, Ure was quick to point out that this was the workers' own fault. At Anderston, near Glasgow, 500 operatives were housed in barracks built by the mill owners. There 'they have been frequently visited by typhus fever of the most malignant and fatal type' simply because they had allowed themselves to get dirty and had not kept the barracks ventilated 'in spite of every remonstrance of the proprietor'. The owner saved them from their own stupidity by running metal pipes from the factory chimney to each of the apartments in the building and, at the end of working day, blowing a blast of air through the whole building. With such benevolence before them, how could they not be grateful? Needless to say, the air blowing did nothing to help, as typhus is transmitted through contaminated water not through the air. There is nothing in the records to indicate how the water was supplied.

There is no need to spend much time on such arguments. Conditions were bad and the rise of the steam mills with their surrounding clusters of jerry-built houses brought a new squalor to the face of Britain. In 1863, the Home Secretary, Sir George Grey, ordered an investigation into the principal towns in 'the cotton manufacturing districts'. A civil engineer, Robert Rawlinson – who had just finished a survey of the plumbing at Windsor Castle – was put in charge. His report spells out in graphic detail just how wretched these town built on the prosperity of an industry:

> Large cotton mills were built, and other buildings connected with the trade of the district, were from time to time constructed, to be surrounded by new streets set out without plan or level, in which houses were built without control or order, the natural surface of the ground being, in many instances, left to form the road ... As there was no main-sewering in such streets, house-draining

Nineteenth-century Manchester: a town of tall chimneys.

was, for the most part, impossible; foul water, and other refuse, consequently added to the evil. Where there was a natural fall, slop and fluid refuse from the houses on the higher ground flowed down and over the surface of adjoining yards at a lower level … This round of sanitary neglect, producing filth, misery, drunkenness, disease, and sometime crime, is as consequent and certain as any other form of cause and effect.

This, then, was the product of 100 years of development, which was said to have set Britain on the road to prosperity.

Faced by such conditions, the textile workers attempted to band together in unions to try and effect some improvement in their lives. Their successes were, at best, limited. They faced hostile employers who were able to call on the law to help them. Fundamental changes seemed to depend on changes in those laws that acted against them. Here the workers had a real problem. Even after the passing of the Reform Act of 1832, they still lacked the vote and so had no direct influence on the Members of Parliament. They had to depend on sympathetic reformers to argue their case for them. In this way at least their position was not so different from that of the plantation slaves – both had to rely on outsiders to make change – and those outsiders tended to be in the minority and faced powerful opposition. At this point, the two stories meet.

Looked at from our position more than a century later, both mill and plantation systems seem intrinsically evil, and so they did to many contemporaries. Were mill owners and planters therefore evil men? Not at all, in the sense that they really believed their actions were justified and justifiable. The mill owner on his part could argue that although he was certainly making a large personal profit, that was his due for providing employment for the many and contributing to the wealth of the nation as a whole. The planter could argue that he had taken in savages and introduced them to the benefits of civilization and religion. Both arguments look suspiciously like special pleading, but they were no different from the rest of us in that they wished to be thought well of by their own communities. They needed to show that they were doing their share to rectify the ills of the world – far more in fact than others were doing. They needed scapegoats – and found them in each other.

If you read through the newspapers, pamphlets, books and journals of the plantation South, you will find two related themes repeated over and over again: conditions on the plantations are idyllic compared with those in the mills of Lancashire; British abolitionists who buy slave-produced cotton are hopelessly hypocritical. The Parliamentary Blue Books, such as the 1833 reports of the Commission on conditions in mills, were seized on with great glee in the South:

> It is shocking beyond endurance to turn over your Records in which the condition of your labouring classes is but too faithfully depicted. Could our slaves see it, they would join us in lynching Abolitionists.
>
> When you look around you how dare you talk to us before the world of slavery? For the condition of your wretched laborers, you and every Britain who is not one of then are responsible before God and Man.

The slave owner reading such tracts could reflect comfortably that his slaves were better off than those unfortunate children in Lancashire. No slave owner, he would proudly claim, would dream, for example, of setting a black child to work before the age of 10. The more thoughtful might have considered that an argument that says because my neighbour beats his wife, I am virtuous because I only beat mine twice a week, lacks validity. They could, however, take comfort from the thought that they need not pay any attention to criticism when the critics are so evidently hypocritical. 'Why beholdest thou the mote that is in thy brother's eye, but considered not the beam in thine own eye?' was a favourite text.

The most famous and popular defence of slavery, which found an honoured place on many a planter's shelf was Elliott's *Cotton is King*. Many of its arguments were directed against the British Anti-Slavery Society. The anti-slavery movement in Britain had been begun with Wilberforce's Society for the Suppression of the Slave Trade. The society had succeeded in getting the trade itself banned in law, but the cynical noted that it had ceased to be profitable when America and West Indies had all the slaves they needed. The society's medallion, designed by Josiah Wedgwood, showed a slave in chains and carried the motto 'Am I not a man and

An American cartoon of 1832 contrasting the misery of Industrial Britain with the 'idyllic' life of the plantation.

a brother?' Elliott pointed out that there seemed to be little that was brotherly in the treatment of factory children. After the trade was banned, the society was reformed as the British and Foreign Anti-Slavery Society. The charge of hypocrisy was reframed. The society had the gall to attack the planter, even though they had a weapon in hand that they refused to use simply because it was against their own financial interests:

> As long as all used their products, so long as the slaveholders found the *per se* doctrine working them no harm, as long as no provision was made for supplying the demand for tropical products by free labour, so long there was no risk in extending the field of operations. Thus, the very things necessary for the overthrow of American slavery, were left undone, while those essential to the prosperity were continued in the most active operation; so that now ... we may say, emphatically, COTTON IS KING, and his enemies are vanquished.

The argument was a powerful one and difficult to refute. There was a strong feeling among American abolitionists that the British found the Americans a convenient target at which to aim social reformers who might otherwise be pointed at objectives nearer home. The Americans in the society joined in the

The Wedgwood anti-slave trade medallion.

attack on their British colleagues, as in this letter from Samuel Watt in March 1843:

> Who are the Slaveholders of America? The Planters? – No! – The overseers, the feculent dregs of society? – No! *The Slaveholders of America are in the City of London!* – in the heart of Great Britain! – excuse me – it is *yourselves*, the people of England who are the real slaveholders of America! – *you* hire him to chain, to whip and to work his slaves to death! – *you* stimulate him by your money! – by your Commerce! – If Great Britain would buy no slave-raised produce, Slavery would not last one year! – That is the plain unvarnished truth, and you will forgive me for telling you so! – many slave-holders go further, and try to excuse themselves by saying Slavery was entailed upon them by England! That it is not their fault 'because had England not introduced Slaves into our country there would be no slavery here!' *then let England abolish it;* by refusing to partake of the profits of slavery – you can do it … *Repeal your Corn Laws!* – repeal all duties upon the products of our *free* states – discriminate between Liberty and Slavery! – *impose a duty upon cotton.*

The Corn Laws were in time repealed, because it suited the interest of the industrialists to have access to cheap food. There was no attempt to impose a tax on cotton imports. This one action was never even contemplated, though the Society on other occasions demanded direct action by the British government. When Texas declared independence from Mexico, they wrote asking the government not to recognise the new state if slavery was permitted. 'The government', they wrote, 'should avail itself of so just and striking an opportunity of using its mighty moral influence'. The mighty moral influence was not brought to bear. Texas was duly recognised and even if it had not been it would have made very little difference to anyone, as Americans of both the North and South had a poor opinion of British morality. They would not refuse to buy cotton from the slave plantations; the charge of hypocrisy was allowed to stand. The South could rest secure in the knowledge that British profits would always be more important than moral issues. British rantings against slavery would have as little effect as American complaints about the British mills. The South and Lancashire were tied together by economic chains too strong ever to be broken by moral qualms. If change was to come it had to come from within the ranks of their own societies.

In Britain, there were reformers anxious to push parliament into a direct involvement in the affairs of the cotton industry. The chief parliamentary exponent of shorter hours, especially for the factory children, was John Fielden. He was one of the leading manufacturers of the day, with a big steam mill in Todmorden on the Lancashire-Yorkshire border. After the passing of the Reform Bill, he was returned as a Radical MP for Oldham, having as his fellow MP the leading radical journalist of the day, William Cobbett. They made a formidable team. Cobbett spoke up for an older Britain, deploring the changes that had reduced the independent yeoman to the role of wage slave. Fielden spoke for the new generation, as the employer with a conscience. He favoured progress, and the new machine age but could not accept that it was necessary to buy prosperity at the expense of little children. Legislation was essential, Fielden argued, because without it a well-meaning employer such as himself was forced to employ and overwork the children or risk being ruined by less scrupulous competitors. His argument was all the more powerful simply because he was a factory owner and not opposing the new developments, simply criticising their misuse. Cobbett made one splendidly sarcastic speech in the House of Commons on 20 July 1833, which began with his laying out all the great institutions that he had supposed had made Britain prosperous, and then thrust in the knife:

But, Sir, we have this night discovered, that the shipping, the land, and the Bank and its credit, are all nothing compared with the labour of three hundred thousand little girls in Lancashire. Aye, when compared with only an eighth part of the labour of those three hundred thousand little girls, if we deduct only two hours a day, away goes the wealth, away goes the capital, away go the resources, the power, and the glory of England.

William Cobbett.

Fielden received enthusiastic support from all the main textile districts, including Yorkshire, where the woollen industry had belatedly followed cotton down the road of mechanisation. Now the other side of the British-American controversy was used in the argument. Again and again it was pointed out that those who were opposing slavery in America were directly responsible for terrible conditions nearer home. The argument was stressed with a flurry of italics in the *Leeds Mercury*:

> The fact is true, thousands of our fellow-creatures and fellow-subjects, both male and female, the miserable inhabitants of a Yorkshire town, are this very moment existing in a state of slavery more horrid than are the victims of that hellish system, *colonial slavery*. These innocent creatures draw out, unpitied, their short but miserable existence in a place famed for its possession of religious zeal, whose inhabitants are foremost in *professing* 'temperance and 'reformation', and are stirring to outrun their neighbours in missionary exertions, and would fain send the Bible to the farthest corners of the globe, ay, in the very place where the anti-slavery fever rages most furiously, her apparent *charity* is not more admired on earth than her real *cruelty* is abhorred in heaven. The very streets that receive the droppings of an 'Anti-Slavery Society' are every morning wet by the tears of the innocent victims of avarice, which are *compelled* not by the cart-whip of the negro slave-driver, but by the dread of the equally appalling thong or strap of the overseer, to hasten, half-dressed, but not *half-fed*, to those magazines of British infantile slavery – *the worsted mills in the town and neighbourhood of Bradford.*

The result of the parliamentary investigations could be read in the Blue Books, which brought out the worst of the conditions and displayed them to the public gaze. Legislation followed, aimed specifically at reducing the hours that children could work and raising the age at which they could start. It was less effective than it might have been, as so few inspectors were appointed that the law could be, and was, evaded. Where employers did keep to the strict letter, the results were not always those envisaged by the law makers. Remnants of the family unit of domestic manufacture had survived into the factory age. Now, with the hours the children worked reduced, but the hours the adults worked unaltered, the family unit was broken. The children worked shifts that no longer coincided with those of their parents and that opened the way to fresh abuses. And all the time the pattern of work itself was changing. More factories meant more operatives were needed, but increasingly they were mainly employing women and children, not adult males. As one contemporary put it in a splendid euphemism, there was 'a diminution in the more expensive class of operatives.'

The legislature gradually brought in new rules, and throughout the nineteenth century there was a general easing of conditions in the textile industry. This was partly due to a greater feeling of ease and comfort among the employers and the country's rulers. When the Chartist movement was at its height, culminating in the General Strike of 1842, the revolution that had convulsed continental Europe was very much in their minds. The mass of the people were lumped together as The Poor, fundamentally inferior but capable of being turned into a dangerous mob. By the 1860s, the threat of violence seemed to have receded, and it became quite respectable at least to talk about giving people a say in the running of their own country, even if nothing very much was actually done. Similarly, the rush to capitalise on new inventions subsided. Employers found little need to make major improvements in the machinery of the mill now that profits had settled down to a steady rate. They felt that they could accept legislation such as the Ten Hour Act of 1847 with few murmurings.

Among the workers there were inevitable changes. By the mid-century few were left who could remember any other way of life and the former domestic system was all but forgotten. But if memories of the past receded, memories of how change had been wrought had bitten deep into the collective consciousness. Benjamin Disraeli spoke of Two Nations, rich and poor, but the divisions went deeper. Even if the economic gap closed slightly, those on each side of the divide saw themselves as opponents not collaborators in a common enterprise. Trust was one of the victims of the industrial revolution. Both side of industry – the very language is that of division – developed their own attitudes. Employers regarded workers as an input that should be obtained as cheaply as possible, bargained over just as they might haggle over the price of raw cotton. Employees regarded a wage increase as something that could never be obtained without a struggle. Improvements in methods and machinery might seem to the outsider to work very much in everyone's interest, but to the antagonists it was all part of a war about pay and conditions. Industrial relations were conducted with bitterness and even, on occasions, violence.

Gradually, however, working conditions could be seen to improve. The changes might have been fought for, but they were real. Society was on the move. Turning back across the Atlantic, the opposite appeared to be true. All was static. The slaves had no possibility of improving their own lives. It was clear that the buyers of slave cotton would do nothing – there was not even a serious attempt to look for alternative sources. But change did come through a bitter and bloody war.

In 1860, Abraham Lincoln was elected President of the United States. The South began to see their economic interest subordinated to those of the North. Two systems were in conflict, with slavery just one obvious mark of those differences. Although the fight against slavery gave the conflict a moral character, it was never at the root of it. In December 1860, South Carolina announced its secession from the Union, and by February 1861 they had been joined by six other states to form the Confederate States of America. Was America to become a loose confederation of states, each maintaining its own laws and identity or an indissoluble union? The issue was to be decided on the battlefield.

Chapter 15

CIVIL WAR

If ever the mutual dependence of the American South and Lancashire needed proving, that proof was provided by the Civil War. As the South fought, Lancashire starved. It was a great turning point in the history of cotton.

As war began, the South immediately found itself faced with a blockade. Cotton shipments – apart from the very few that managed to break through – virtually ended. The Confederacy had high hopes of British intervention on their side, as the effects were felt in the mills. Southern protagonists set out their case to the British, largely through a pro-Confederate newspaper, *Index*. Their hopes soon collapsed. War can mean profits as well as losses. The effect on cotton may have been disastrous, but elsewhere trade boomed. British marine trade, which had been feeling the competition, prospered. Armament manufacturers and the metal trades as a whole had all the work they could handle. Even the textile manufacturers were able to salvage some profit, by raising the price on the stocks in hand and on whatever cotton they could get from other sources. The unhappy operatives, however, had no other resources to fall back on: they felt the full weight of the disaster. Just consider these statistics for the Lancashire industry in November 1861 and the same week in 1862:

1861 1862
Average weekly consumption of cotton (400-lb bales) 49,000 18,000
Operatives working full time 583,950 121,129
Operatives working short-time 165,500
Operatives out of work 247,230
Estimated loss of wages (weekly) £169,744

Here is a story of real hardship, but Britain was not about to go to war on behalf of unemployed cotton workers, even if there were a quarter of a million of them. There was, in any case, very little sympathy for the Southern cause, except among the anti-democratic elements of society, and quite a lot of support for the North. The different attitudes were expressed in popular ballads of the day. This one indignantly denounced the Unionists who had attacked a British merchantman attempting to break the blockade:

What did the valiant Yankee mean by manner so offensive
To stop our craft upon the sea that had no force defensive?

The defence of Fort Sumpter where the Civil War began.

I tell you why: the quarrel for the everlasting Nigger
Made Northern states 'gainst Southern states pull fratricidal trigger.
The Southerners had hated long the Northern knavery
Which spurn'd the name, but lov'd the gain, of woolly slavery.

A rather different view was given by *A Factory Girl*:

… hence it is that we
Lack cotton to employ our industry;
And cotton failing, causes work to fail,
And labour is the poor man's capital.
To be deprived of labour is to be
Plunged in the depths of want and poverty.

But can it be that free-born Britains have
Depended on the labour of the slave?
Yes! So supplies of cotton were assured.
They cared but little how they were procured.
So they had cotton, cared not tho' it were
Stain'd with the blood of slaves, nor did they care.
Tho' on the hands that pick'd there should be
The galling chains of hateful slavery.

If this was in any way typical of attitudes among the unemployed of Lancashire then they were quite astonishing in their selflessness, for their suffering was great. A visitor from London toured the district recording what he saw. In Stockport, for example, he found a woman living in one of the mean, single-storeyed courtyard houses. Everything had been sold except the bed and its covering and her cooking pots. Her five children were with her, but not her husband. He had obtained 9s and 11d worth of provisions on credit, and when he was unable to pay the debt was thrown in gaol. The visitor asked her if she did not get any relief:

Yes, sir, I do, and very thankful I am for it; but I only have 3s 6d a week, and what is that? In good times my master used to make £1 to £1 5s a week, and then we thought we could only just live but now see what we have come to!

Everywhere he went he found the most appalling poverty, yet the operatives he visited answered his questions, however impertinent, politely and cheerfully. At one house he was received kindly and only afterwards did he learn that the woman had just returned from burying her child. 'Do you think they are sustained in their trials by dependence upon Providence?' he asked in amazement. 'Or does their resignation arise from sheer insensibility?' Fortunately, he had left the district before he posed that question. He found the same poverty everywhere. Some families had sold everything and moved in to join friends in already overcrowded houses to save rent. Many were helped by local traders, who allowed almost unlimited credit. Others, especially butchers, simply went out of business themselves and began sinking towards the poverty of their former customers.

There was much bitterness over the failure of the mill owners to help. In wealthy Preston, for example, less than £2,000 was raised for relief funds, and only forty-eight of the seventy-one mills contributed anything at all. The greatest bitterness, however, was reserved for the Labour Test. The poor who applied for relief had to show their willingness to work. Men who had been indoor workers, badly clothed and close to starvation were sent out in mid-winter for jobs such as stone breaking. It was not merely cruel but often fatal.

Sermons were preached in the churches and chapels, although some offered little more than pious platitudes. The preacher who began his sermon with 'Everyone has heard of the terrible distress which God in his infinite wisdom has allowed to fall on the manufacturing population' can have done little to restore the faith of the poor in divine benevolence. Others were more practical, demanding national relied for the sorrows of Lancashire:

> The operatives suffer then in consequence of a national policy; therefore the relief of that suffering should also be national. Not, perhaps, if they had denounced that policy – if they had raised disturbances on account of it; if they had said, 'it is better to violate an international law of England than that we should starve.' But they have not said so: they remained quiet; they have even fully concurred in the policy with an unanimity that is astonishing. The Government of the country owes a deep debt of gratitude to Lancashire.

The preacher of the above sermon, published in *Distress in Lancashire* in 1862 looked at the fortitude of the unemployed and did not see insensibility. He summed up his attitude with one fine, ringing phrase. 'To Lancashire we should give with pride, as to one who has honoured us, and who was noble in ruin.'

Public relief handouts for unemployed Lancashire cotton workers.

There was some relief offered by the central government, but they still insisted it was tied to the need to make work. Robert Rawlinson, the civil engineer, was given the job of seeing what could be done in the way of public works – paving streets, laying drains and creating public parks. It was a cumbersome system. Rawlinson had to make his findings known to the local authorities, who then had to apply for a government loan and then, if that was granted, the business of relieving poverty could begin. Even then it came a long way short of what was required. Because of the insistence of tying relief to work, of the estimated £1.5 million being made available, about £1 million had to be spent on materials. But, as Rawlinson pointed out, the work generated employment among suppliers and subsidiary trades. And at least the work did have a useful end product – unlike most of the work handed out under the Poor Law. In a report of November 1864, he estimated that there were over 6,000 directly employed and another 2,000 finding subsidiary work. It was something, but set against the 250,000 unemployed, it was precious little. Private charity helped to pad out the meagre government relief, but what was really needed was a return to work, and that work needed cotton. Lancashire had discovered the truth of what some had been saying for a long time, that it was dangerous to rely on a single source. The Cotton Supply Association was set up and began to cast about to find alternative suppliers.

There was only one obvious direction for Britain to turn: eastward back to the origins of the cotton trade – to India. One pamphlet by Hensry Ashworth, produced at the beginning of the war, managed with great economy to incorporate most of its arguments into its title: *Be just to India; Prevent Famine and Cherish Commerce*. Old arguments were rehearsed and there was a now familiar condemnation of the policy that had allowed the supply of Indian cotton to fall to less than 10 per cent of the total used in Britain. Why did India not produce more? Everyone had an answer. Mr Mangles, a former Chairman of the East India Company, had his firm idea on the subject: 'I have made the largest admissions with reference to the want of roads, which I say, is the only real obstacle to the exportation of cotton in large quantities from India.' There was certainly ample evidence of bad roads. An Indian civil servant described a journey of 12 miles that took seven hours of continual, painful jolting. 'On his way, the *mamlatdar* amused us with several stories of accidents which had occurred on this road, one of which is related to the sad fate of a *banian*, or trader, who received such a jolt as to make him inadvertently bite the end of his tongue off.'

Some improvements were made in roads, and new roads were designed and built specifically to serve the cotton growing regions, but the work went on slowly and fitfully and made little real impact on India's transport problems. The answer was widely felt not to lie with these roads but with railroads. The first great protagonist for railway construction was Lord Dalhousie, Governor General from 1848 to 1856. His dream was not primarily concerned with trade, but rather he planned a network of trunk routes to join the cities of the coast to those of the interior. The plan was to be broad in concept and broad in gauge, for Dalhousie took evidence from Britain on the relative merits of the Stephenson 4ft 8½in gauge with Brunel's 7ft and opted for a compromise 5ft 6in gauge. The first companies, the East India Railway and the Indian Peninsula Railway, were formed in 1845, capital being raised on a curious guarantee system. The government gave investors a guaranteed return of 3 per cent (which was soon raised to 5 per cent) on their investment, regardless of whether the railways made a profit or a loss. Furthermore, the original investors had what amounted to a cash-back guarantee should they choose to sell up. As the government could have borrowed the money from the banks at a lower rate of interest – and as none of the railways were to make a profit for some time – this was good value for shareholders but a pretty poor scheme for the government. Nevertheless, it was a start and a certain amount of money came from Lancashire investors, encouraged by the failure of the cotton crop in America in 1846. Dalhousie himself was well aware where the Lancashire investors' interest lay. He wrote in April 1853: 'I know that the British plutocracy intend to endow India with railway with the exclusive view of extracting at diminished expense the cotton and other raw materials of their manufactures.'

Inevitably, the Indian Mutiny brought railway construction to a temporary halt, and work resumed without the inducement of the guaranteed return, the funds

The first train running on India's first railway from Bombay to Thana in 1853.

were notable less forthcoming. A Bombay merchant expressed his disgust in the *Daily News* in May 1861:

> It is a remarkable fact that though the Manchester interest has been constantly urging the government the duty of their developing the country, it has nevertheless entirely withheld its pecuniary support from the railway undertakings which were the first and most important step towards this object.

Yet the railways were built, and the port of Bombay was connected to the cotton fields of the interior. This was an immense undertaking, for it involved pushing a line through the massive range of cliffs and ravines of the Western Ghats. Lord Elphinstone had prophesied that the passage of the Ghats would not be achieved without many casualties.

> Every possible means must be taken to lessen the risk – but it would be idle to expect that … we should overcome the physical difficulties which we have to confront in making railroads in such a country as India without heavy sacrifice of human life.

His words were to prove all too well founded. The men faced difficult terrain, attacks from wild animals and, in the rainy season, atrocious conditions. In the monsoons of 1859–60, all work was halted among the 30,000 labourers, as cholera struck at some 10,000 of their number. But the work was completed by 1865. Further progress was to come with the realisation that the railways could help to provide aid for the regular famines that afflicted India in the 1870s. To save money, the new lines were built to the narrower metre gauge.

Other critics of India's cotton production looked at the shortcomings of the East India Company, who collected rents from the land but failed to use the money where it was most needed, in irrigation. In fact, they had even made matters worse by failing to maintain the systems already in place. In Poona, for example, during the years 1849–51, £185 was set aside out of a revenue of £80,500; in Belguam, £75 out of £125,000 and in Sholapur, nothing at all was spent.

The response of the government was immediate, though actual development was slow. As the role of the East India Company in governing India came to an end in 1858, a new irrigation programme was begun with the construction of the Sirhind Canal in the Punjab. This was the beginning of a canal network that was to turn a semi-desert into a rich cotton-growing region. But such schemes take time to develop, and there was still the urgent problem to be solved – how to get more cotton from India to Britain, and cotton of the right quality.

Most commentators agreed that the root of the quality problem lay with the land tenure system. The cultivators, the *ryots*, were permanently in debt to the middlemen, the *usurers*, and both were preoccupied to find ready cash to pay

The dockside in Bombay with cotton bales piled high, c.1900.

British taxes. So the old difficulties remained. The cotton that was sent to Britain was poor quality, adulterated with poorer. The Cotton Fraud Act of 1863 imposed heavy penalties on anyone caught adulterating cotton or using unlicensed gins, but as the choice lay between possible prosecution under the Act or certain prosecution for non-payment of taxes, it made little difference. It was not this sort of legislation that was needed, but a whole change of attitude. In 1856 an anonymous 'Indian Civil Servant' argued that what was needed was long-term investment:

> The immense field that is open for the employment of European capital in India has never been conceived by capitalists at home. There are fortunes to be made in India with far greater facility than can be commanded in a country where every profession and every trade is overstocked. Without competing or attempting to compete with the native producer of the raw material, it would make a fortune of any man who, with a few thousand pounds of capital, would set up improved steam-worked machinery wherewith to clean cotton thoroughly up the country, and to screw it into bales for shipment to England at once.

Hand picking cotton in India.

As civil war in America became at first a threat and then a reality, so the clamour for investment in India became more insistent. The Manchester Cotton Company was formed with plans to raise £1 million; by July 1862 they had just a little over £40,000. But they were able to collect enough cash to buy cleaning and pressing machinery which was duly despatched. Unfortunately, a promised new pier and road in India were scarcely begun, and the expensive machinery was left to rust on the beach. The scheme came to nothing – not one pound of cotton was ever brought to Lancashire by the company.

Yet there was a genuine surge of interest in extending and improving India's role, encouraged by the exorbitantly high prices that could be obtained throughout the Civil War years. *The Times of India* was crammed with advertisements for ginning machinery of which this, of 23 April 1862, is typical.

All persons interested in machinery for separating cotton from the seed are invited to inspect the Patent Improved machinery for that purpose by Platte Brothers & Co., to be seen at the office of the undersigned, who is prepared to receive orders for the same.

The orders came in and venerable Platt gins are still found at work in India.

Nineteenth-century cotton gins, still in use at Karjan, Gujarat.

The Indian growers began a period of unprecedented prosperity and, in spite of all the shortcomings, were able to do a great deal to plug the gap left by American cotton. Average imports of Indian cotton into Britain for the five years from 1855–60 ran at 192 million lbs per annum; in the next five years the figure more than doubled to 430 million lbs. But in spite of all the successes that could be shown, there was still a lack of faith in the long-term prospects for India. The Manchester men continued to see importing from India as a stop-gap measure. The war, they argued could not last for long, and once it was over things would be as they were before and old relationships would be resumed. They were, in fact, entirely mistaken.

Chapter 16

AFTERMATH

No one in the South ever doubted that defeat would mean changes that would encompass far more than the end of slavery. It had been a bloody and costly conflict and there was a price to be paid by a region already devastated by the passage of war, for the war itself had been fought almost entirely in the South: great cities such as Atlanta and Charleston had been laid waste; plantations had been destroyed; bales used as barricades and cotton left to rot in the fields. A cotton tax had been imposed, and government agents sent to the South to collect it. They were empowered to buy and sell cotton on a commission basis, so as to pass on the tax. Many bought the cotton for themselves at artificially low prices, passed on the tax, then resold at the true market value, pocketing the difference. Secretary Hugh McCullough in Washington remarked, 'I am sure I sent *some* honest agents south, but it sometimes seems very doubtful whether any of them remained honest very long.' The tax was estimated to have cost the planters some $68 million. The carpetbaggers who invaded the South in the years immediately following the war did little to help, nor did those Northerners who

Charleston in ruins at the end of the Civil War.

bought up plantations in the belief that they had only to sit on a veranda with a mint julep, counting the profits.

There were numerous tasks facing the Southerners who wished to rebuild the plantation system, and they often began from a very insecure financial base. Many of them had committed themselves to the war effort, and now all they had left was worthless Confederate money and equally worthless Confederate bonds. Worst of all, in Southern eyes, the greatest capital asset that the plantations had possessed was removed from them overnight. The slaves were free. To the South, millions of dollars' worth of property had been taken from them with no compensation whatsoever. To the North it was simply a case of the South being forced to acknowledge that human beings could never be property to be bought and sold. To have paid compensation would have been to accept that a human had a cash value. It was inevitable that the change would be made, but only now did it become plain just how far the system had depended on the trading in slaves for its viability.

All these problems, great as they were, had recognisable dimensions. Debts could be paid, and the land was still there. Given time, crops would grow again and prosperity return, however slowly. Other problems were less tractable. The South had built a mythology of the 'faithful slave'. The war had revealed it for what it was. Slaves had 'deserted to the enemy' leaving planters feeling both bewildered and betrayed as Louis Manigault noted in his diary:

> With us upon Savannah River, my favourite Board Hand, a Man who had rowed me to and from Savannah from my earliest recollections of Gowrie plantation ... a Negro we had all esteemed highly. Singular to say, this man 'Hector' was the very first to murmur, and would have hastened to the embrace of the Northern Brethren, could he have seen the least prospect of a successful escape.

The planters who had prided themselves on their paternal attitude towards their slaves now, in their rage, often turned them away penniless with nothing but the clothes on their backs. Other blacks took the decision for themselves. Not surprisingly, they headed off to the towns, eager to enjoy a taste of freedom. They were in no great rush to exchange slavery for any new kind of subjection to their former masters. But however reluctant the two sides were to renew acquaintance, some sort of relationship had to be cobbled together. Slave owner and slave had to adapt to new roles as employer and employee.

The old plantation could not simply be converted into a new model. The most obvious system to move towards was one in which hands were paid wages, but there was a real difficulty to be overcome – the shortage of cash. Money wages in the South as a whole plummeted from an average of nearly $140 a year in 1860 to $100 in 1868, and in many cases there was no money at all until the crops were sold. Owners offered contracts to their labourers, specifying payment at the end of the season but the latter had little faith in their former masters' promises and no understanding of legal documents. Many plantations developed a system whereby gangs of workers under black supervisors were given the means to live, occupied the former slave quarters and were offered a share in the crops. But this was too redolent of the old days and old ways. Gangs broke up into family units and the families dismantled the old shacks and rebuilt them close to their own particular part of the plantation. The monumental structure of house and quarters and surrounding fields was being broken apart. The South was becoming a region of landlords and tenants.

The tenants could be divided into three categories. At the top of the ladder were those who paid a regular cash rent for the land they farmed, providing all the necessities for the work from their own funds and living on whatever profits they made. Slightly lower down the scale were those who, unable to raise the capital for a cash rent, paid their way by handing over a portion of their crop to the landlord. At the bottom were the former slaves who started with nothing – no home, no equipment, no seed and no money. Here the owners supplied the materials for farming and kept the families provided with the necessities of life until the harvest came round and the crops were sold. A share of the crops was handed over as rent and the year's accumulation of debt often accounted for the rest. In many cases, when the entire crop had gone, some part of the debt still remained unpaid. The tenant started the new year, as he had the last, in debt and the cycle began all over again. There was no escape from this ring of debt, and those who ran away leaving it unpaid were hunted down and were fortunate of they escaped a lynching. In effect, the landlord was hiring labourers and paying them in kind instead of cash, while binding them to their holdings as firmly as slavery had done.

The black tenants were often given the very best land, such as that of the Mississippi Delta, yet often made little of it. With so little incentive, few prospered and crop yields were low. But it suited the owners well enough. They were getting work done at rock bottom prices, and so they favoured the black tenants. No one else, as one planter noted, 'would be as cheerful or so contented on four pounds of meat, and a peck of meal a week, in a little log-cabin 14x16 feet, with cracks in it large enough to afford free passage to a large sized cat.'

The cheerful victims of this new system were the share-croppers, who were to typify one aspect of this new South. Not all share-croppers were black, for there was still a poor white population. But, as in slavery days, they could not be brought to work under the same conditions as the blacks. They were left to farm the poorest lands, scratching what they could out of the thin soils of the hills, and only working on the richer lands of the plain when they could obtain work with a suitable status, such as overseer on one of the large plantations. Poor white and poor black were divided by the colour of their skins, though united in poverty.

A share-cropper's cabin.

By 1880, cotton production was back to pre-war levels, even though society had gone through a fundamental revolution. Changes were also felt in the manufacturing sphere. Many Northern mill owners had been prepared for the war, having bought heavily in the immediate pre-war years and having arranged for their cotton supplies to arrive by railroad rather than through the Gulf ports. Nevertheless, there was cotton famine in New England, just as there had been in Lancashire. The more prudent mill owners did what they could to keep their businesses going. Some turned to spinning cotton waste to make a coarse thread and found it profitable. Others used the time to modernise their mills. Some simply closed up shop. In April 1861, the directors of the Merrimack Manufacturing Company of Lowell sold off their cotton stock, it must be said at a handsome profit, to other New England mills. Ten thousand operatives were simply turned out onto the streets. Charles Cowley, the historian of Lowell, writing shortly after the war, had no doubts of the implications:

> This crime, this worse than crime, this blunder, entailed its own punishment – as all crimes do by the immutable laws of God. When these companies resumed operations, their former skilled operatives were dispersed, and could no more be recalled than the Ten Lost Tribes of Israel. Their places were filled by the less skilled operatives whom the companies now had to employ. So

serious was this blunder, that the smallest of the companies would have done wisely, had they sacrificed a hundred thousand dollars, rather than thus lose their accustomed help.

For Lowell, it marked the final stop in the story of the mill girls. When the mills reopened, the immigrant population filled every vacancy.

At first it seemed that there would soon be a return to normality and the New England manufacturers began to expand as the flow of cotton was resumed. New mills were built and the spinning capacity in the North went up 12 per cent between 1868 and 1870. But other manufacturers began to look South. There had been a small start in the South before the war (see Chapter 9) but it had always proved difficult to run a factory system within the context of a slave economy. Now that obstacle had been removed and there was a lot to be said for mill construction in the South. First there was the social argument. There was a major problem in finding employment for the poor whites, who could not be persuaded to take on any farm work that gave them the same status as black workers. Mill building was also seen as vital to the whole process of Southern reconstruction, designed to integrate the region into the main framework of American life. This view was forcibly expressed by Francis W. Dawson who wrote in 1880 that bringing the whites to mill work introduced them to 'elevating social influences, encourages them to seek education, and improves them in every conceivable respect'. He also boasted of the economic gains to the community, quoting figures to show that South Carolina had 2,296 operatives, upon whom 7,913 persons were dependent for support, from a monthly wage bill of $38,034. As this represents a monthly wage of just over $16 per worker, or if you add in all those dependents, less than $1 per person per week, it does not look overly generous. Dawson also reported that company profits were running at between 18 to 25.5 per cent per annum. Whether philanthropy or profit as a motive for those who invested in the South is a matter of speculation. But some of those who moved South, such as Lockwood Greene & Co., were quite honest in describing their motives:

> As compared to New England and the Northeastern part of the country, the South has the advantage of longer hours of labor, lower wage scales, lower taxes, and legislation which gives manufacturing plant a wider latitude than is usually possible in the North in the way of running over-time and at night … the South is … fortunate in having a supply of native American labor which is still satisfied to work at low wage.

The organisation of Southern mills very much followed the pattern established in the ante-bellum years, a pattern of paternalism. The company would provide – but in providing they came close to the old truck system that had obtained in some British industries in the previous century. Mill hands were tied to

An idealised view of the new Southern cotton industry in the post-war years.

their jobs by the bills they ran up at the mill commissary that provided them with the essential of life on tick, as tightly as the share-croppers were tied to their holdings. The programme succeeded only because it existed against a background of general poverty. But succeed it did. Families came down from the hills to camp out on the site of a new mill, waiting for jobs to be supplied and company housing to be built. In 1860 there had been some 10,000 employees in Southern mills, but by 1890 the number had quadrupled. The conditions for growth were improving.

America was expanding. The railways were thrusting out from the east and the west coasts, and by 1860 the transcontinental route was completed. New floods of immigrants came in, reaching half a million a year by 1873, and they were all new customers needing to be clothed. And the opening up of the West also opened a route to the Far East and new trade opportunities. It seemed to make less and less sense to send cotton to Lancashire when there was rapidly growing demand for cotton in the new mills of the South. A new society was being built there and though the process would be long and often painful, there was no return to the old ways.

Britain was slow to recognise the threat to her old dominance, just as manufacturers were slow to recognise the need to improve on the machinery developed during the first rush of inventive energy in the late eighteenth century. Cotton spinning was still being divided between the throstle, an adaptation of Arkwright's water frame for the steam age, and the mule. They had worked well in the past, continued to work well and had made profits for their owners. So where was the need for change? There was little excitement when John Thorp of Providence, Rhode Island took out a patent for a new spinning machine, the ring frame. This used rollers to extend the yarn but twist was now delivered and the thread wound on by a 'traveller' moving round a ring set over the bobbin. It was a continuous process for stretching and winding on the yarn and faster than previous machines. True, it took some time to perfect it and the thread produced was slightly inferior to that from the mules, but the British lack of interest suggested they had lost their enthusiasm for innovation. This lack of enterprise was one factor in the industry's decline: the other was the loss of old markets.

Once the Civil War was over, Britain dropped India as quickly as she had taken her up at the beginning of the cotton famine. Rather than seeing the war years as grim evidence of the effects of overdue reliance on one source of raw material under the control of a foreign power, they chose to see it as little more than an unfortunate interruption in the natural flow of events. India had served its purpose and now things could get back to normal. The strings of empire could again be pulled to make India dance to a British tune. The puppet, however, refused to perform – it had a will of its own.

It was not immediately apparent that a decisive shift had occurred. The signs suggested quite the opposite. The price of Indian cotton tumbled as American supplies were resumed and by 1872 exports to Britain were down to two thirds of what they had been in the war years. Fortunes had been made in those years, the like of which had never been seen before in the Indian cotton trade. As always in such circumstances, those who used their quick gains foolishly soon fell back

A little girl operative stands between rows of ring-spinning machines in a South Carolina mill.

into bankruptcy, while those who invested wisely for the future prospered. The decline in sales to Lancashire did not mean that there were no other markets available. India was no longer to be the major supplier to Britain, but Britain no longer dominated manufacturing as she had done at the beginning of the century. The rest of Europe was developing an ever-larger part in textile manufacture, and with the opening of the Suez Canal in 1869, transport costs were greatly reduced. Harry Rivett-Carnac, the Cotton Commissioner for the Central Provinces, noted the trend in his report for 1868–9:

> I have been very much struck with the direct trade in cotton between India and the Continent. Last year a French house in Bombay headed the list of shippers from that port. This year the number of foreign mercantile houses has largely increased. A French house has purchased land, and set up full presses in the Berars, and there appears t be a determination in France to deal direct with India for her cotton.

Other European countries were also increasing their trade, and a new customer appeared in the East that was to take an ever-increasing share in the world of cotton – Japan.

Once the shock of the price collapse had died away, there was more or less a return to normality in the Indian cotton fields, but with one important difference. The various works of improvement in transport and irrigation that had begun in the war years were not allowed simply to grind to a halt. India might not be able to compete on equal terms with America, but things were improving. There remained, however, the social system that kept the peasant farmer permanently in debt, unable to make the necessary improvements to bring higher yields and better quality. Authority seemed to feel it was none of their business and would probably have continued to ignore the problem if the peasants hadn't taken matters into their own hands. In 1875 there was a widespread uprising against the money lenders; promissory notes and mortgages were simply taken away and destroyed. The revolt was soon over but it forced the government to take notice. New and better ways of financing the poor farmers were clearly necessary, and one likely method was through the formation of farmers' co-operatives, which could raise funds to provide credit to the members. Deliberations were not speedy and the first Co-operative Credit Societies Act was only passed in 1904. It was a modest beginning but was to have a profound effect.

Money had flowed into India during the American Civil War years, and though some was frittered on useless schemes, funds also went into building cotton mills. Such mills were not altogether new to India. There had been earlier attempts to start the industry – though not very successfully – in Pondicherry and Calcutta. Real opportunities came when the industry became established on the west coast. One of the earliest entrepreneurs to try to start up the industry was a merchant,

Ranchhodal Chatalal. He had the money to invest but no knowledge of the industry as he candidly admitted when he wrote to England in 1848 on behalf of his consortium, asking for quotes for equipping and supervising a mill:

> These native gentlemen being totally unacquainted with machinery, either theoretically or practically, it becomes a matter of the greatest importance to them, when embarking on such an undertaking, that they should possess a guarantee that the machinery supplied will perform all that is stated to be capable to do.

That plan came to nothing, but at much the same time an Englishman, James Landon, had arrived in India to take over the running of the Cotton Experiments Centre, which the East India Company had established at Broach. He went on to make a fortune and ran his own steam-powered gin. In 1851, he was approached by Ranchhodal, who proposed a partnership to establish a mill at Broach to spin cotton from the surrounding districts of Gujarat. They failed to agree terms and eventually Landon went on alone to establish the Broach mill that began work in 1855. Ranchhodal meanwhile went on to found the basis of what would become the thriving mill industry of Ahmedabad, whilst the Parsee merchant Cowarjee built the first of what would become the even more important mill industry of Bombay.

The mill industry was already established in a modest way when investors began looking for new opportunities in the 1870s. There was money available and with less cotton now going to Britain, the raw material was there in abundance. The period saw a boom in mill construction in Bombay. In 1870, there were eighteen mills in the area, by 1875 there were thirty-six and

Sampling cotton from bales stacked in the warehouse c.1870 by J.L. Kipling.

by the end of the decade the number had risen to forty-two. By then, cotton manufacturing had become India's biggest industry, employing over 40,000 workers.

From the start, India faced hostility from Manchester. British textiles came into India on very low tariffs – a source of constant complaint from the Indian authorities who suffered from a chronic shortage of revenue. Indian textiles, on the other hand, were heavily taxed when exported to Britain. Attempts to change the situation made one thing very clear: India was being ruled for the benefit of the British not the Indians. Sometimes this fact was hidden, but some were prepared to admit it with complete frankness. Sir John Strachey, the Minister of Finance, had this to say in his financial statement of 15 November 1877:

> We are often told that it is the duty of the Government of India to think of Indian interests alone, and if the interests of Manchester suffer, it is no interest of ours. For my part I utterly repudiate such doctrines. I have not ceased to be an Englishman because I have passed the greater part of my life in India and have been a member of an Indian Government … though I have duties in India there is no higher duty in my opinion, than that which I owe to my country.

If that was the opinion of a member of the Indian government, then there was even less possibility of a sympathetic hearing from Westminster.

Nevertheless, India continued to make headway, largely through an intensive programme of modernisation. In 1883, the first experiments were made with the

Power looms in a mill in Baroda, Gujarat.

use of ring frames – a development still largely ignored in Britain. The results were highly satisfactory and orders were sent to the most famous machine manufacturer in Lancashire, Platts. They refused to supply the Indian market, but Howard and Bullough of Accrington proved more accommodating.

This was inevitably regarded with extreme distaste in Lancashire, where the mill owners who had fought their workers' attempts to improve their condition through legislation, now developed a moral conscience. They railed against the poor working conditions and low pay of Indian mill hands, without mentioning the fact that their counterparts in Bombay were undercutting them. An official enquiry into conditions in Bombay did reveal that they were little better than those that had prevailed in the darkest days of the Industrial Revolution in Britain. Here is James Cocker, a mill owner from Bombay.

Of course there is no lighting up in Bombay. There is no gas. The daylight there is very little different from what it is here, winter and summer taken together, but it is more equally divided. The consequence is that we start our mills in the height of summer, in the long days, about a quarter-past five in the morning, and close about ten minutes to seven at night.

Do you stop the mill?
Yes. It is called half an hour, but in a great many cases it is reduced to twenty minutes, or even a quarter of an hour. The Factory Act is null and void here.

Is that for the day?
Yes. The hands there have no system of regular meals, as we have. They get their breakfast all hours, from seven to eleven, while dinner lasts from one till four o'clock. You will never go through the mill without finding some one, behind the mule gate, or in the carding room, eating … In winter time we commence about twenty minutes past six in the morning, and we run until twenty to six at night. I am giving you the longest and shortest time. The average working day is 12¼ hours, Sundays as well. We reckon to stop every other Sunday for cleaning purposes, but that is not a hard and fast rule – far from it. To explain myself: supposing, we will say, that the engine has something to do with it any day of the week prior to the Stopping Sunday – say it breaks down on Thursday – and there would be a necessary stoppage to repair it, the consequence is that instead of stopping two hours only, they would stop half the day and set the hands to clean working all the next Sunday. We work about 80 hours per week.

What holidays to have you beside these odd Sundays?
Nine full days in Bombay and two half days. I have made a kind of diary … We should, as I said, have stopped every other Sunday, but, as showing the extent to which we actually do it, we only stopped eleven Sundays during 1887, and then

three out of the eleven were stopped simply because we were forced to stop, because the engine broke down.

It would seem that India was simply repeating what had happened in the early years of industrialisation in Britain. Things did change, however, but slowly. And real progress only came later in the twentieth century.

So, while India was changing, what was happening in Britain? The short answer is – not a great deal. Although manufacturers were aware the world was changing, they could still rely on the system of imperial preference to sell their goods abroad. Britain had once relied on the lucrative markets of Europe and America to sell their yarn and cloth, she now depended on markets in the Far East, particularly in India. In 1820, over 60 per cent of manufactured cotton went to Europe and America, while in 1880, that had dropped to less than 10 per cent. Exports to the Far East went up from 32 to 82 per cent in the same period. Industry might appear healthy enough, with decent profits, but in reality rot was already nibbling away at the fabric. Inefficiency was protected by empire, but if the empire collapsed nothing would stop the rot from spreading. The American Civil War had changed the South forever; it would never again be a one-crop purely agrarian society and India would no longer be content to accept a role decided by men in Manchester.

Chapter 17

FULL CIRCLE

By the latter part of the twentieth century, the cotton industry of the world had become a truly international business, no longer dominated by any one country or even any particular region. Mills were equipped with new machinery developed by different companies in different countries. Once all the new ideas had been British; this ceased to be true in the nineteenth century when the Americans introduced ring spinning. If we look at two different machines, we can see a pattern emerging. A British inventor, A.W. Metcalf, actually worked out a new spinning method, which was patented in 1901, and was very similar to that used in many mills today – open-end spinning. This involves blowing the fibre along an airstream towards a very rapidly rotating cup. The fibres are laid in a groove

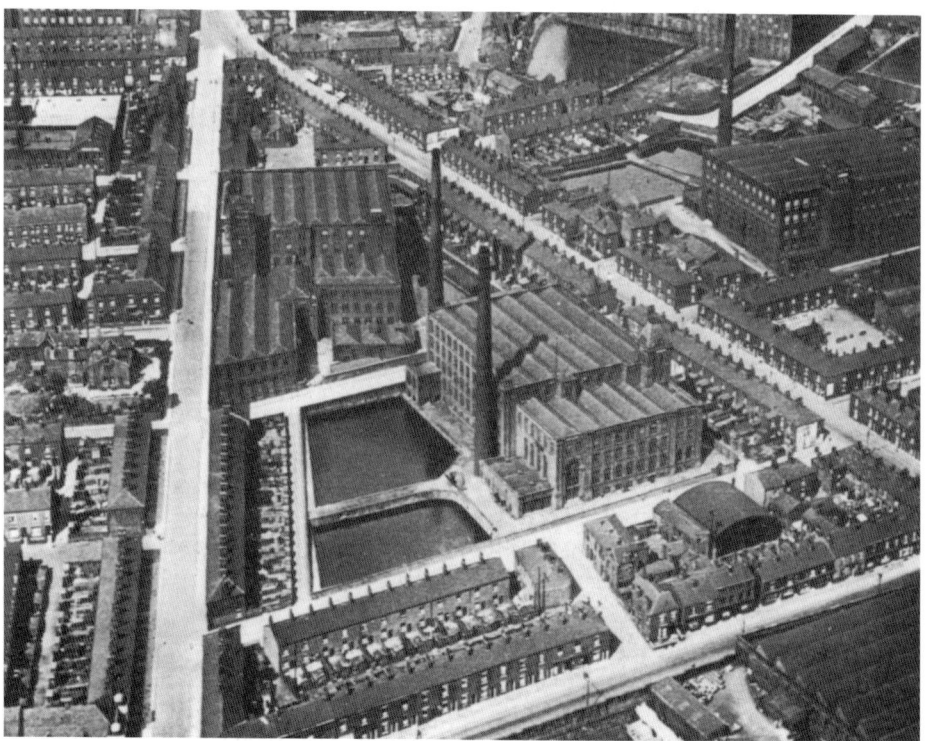

Oldham in the 1930s.

of the rotor, and as they are spun off, they are twisted together. It is an efficient machine, yet nothing was done with Metcalf's patent until 1960 when open-end spinning was introduced, not in Britain, but in what was then Czechoslovakia. In weaving, it had long been recognised that a lot of time and effort went into pushing a bulky shuttle backwards and forwards across the loom, and still more time was wasted replenishing the shuttle with weft. Two new systems were introduced that do away with the shuttle altogether and use a continuous supply of weft. In the first the thread is shot across the loom by a 'bullet' or 'gripper' which collects the yarn. The alternative uses 'rapiers' which shoot out from either side of the loom, one carrying the thread, which is passed to the other when they meet in the middle. Development of these two types took place in America and Switzerland.

The British cotton industry was winding down throughout the last century. It might not have appeared so at the start when British mills dominated the world market, but it became ever more apparent as time went on. Look at a photo of any mill town at the start of the century and you will see a forest of chimneys, each sending its contribution to the pall of smoke that hung over the town. A photo taken at the end of the century would have shown a very different picture, the chimneys either gone or left as monuments to a dying industry.

Oldham in the 1980s with not a mill in sight.

The British cotton industry has always been an exporting industry. In the years before the First World War, the country was producing 8,000 million yards of cloth a year of which 6,900 yards were exported, but everywhere the old markets were being assailed. The monopoly of the early years which had given Britain the lead was ended, and other countries were not only supplying their own needs but exporting as well. They were doing so, what is more, using techniques and machinery that were often up to date and efficient. The British industry was, to a marked degree, guilty of smug complacency. It had always prospered, was prospering and would continue to prosper. In the early 1920s, the years of optimism following the end of the war, a great deal of money went into the cotton industry – not as in other countries in development and improvement but in takeovers. There appeared to be a curious alteration in thinking between 'big is best' and 'small is beautiful'. These were the years of big investors taking over small firms in order to form 'more efficient units'. Unfortunately, the extra efficiency that was expected seldom materialised. The investors, who often knew little about the industry, failed to appreciate the fragmentary nature of much of it. The small firms had often been run by individuals in their own particular way to find their own way of finding a profit. Lumping a dozen such concerns together simply produced a hotch-potch of different patterns of work. And all the time the opposition was getting stronger. Italy was taking an increased share of the Balkan market – helped by taking up the surplus of raw cotton from India that Britain had rejected. America was looking to its own factories to supply its needs and was changing from being an importer of cloth to an exporter, especially finding a market in what had been traditionally British across their border in Canada. In the Far East, the fastest growing industry was in Japan, while India was taking less and less from Britain. Just to add to the problems facing the cotton trade, there was a new competitor, the man-made fibre. The first successful fibres had been developed in the 1890s and by 1904, Courtaulds, who had done much of the research and development, took over the manufacturing rights for rayon.

All these factors demanded a vigorous response if Britain was to continue to hold a dominant position in the textile world. That response was not forthcoming, not necessarily because of a lack of initiative from the manufacturers, but for a whole variety of reasons, including government economic policy. In 1919, Britain was still a leader, if only just, but as the demand for cotton kept growing, Britain's share of the world market kept declining. Before the First World War, the country had held over two thirds of the market; by 1925 it was down to just half, and still falling. Then in the 1930s, everything began to fall apart with the great depression that followed the Wall Street Crash.

Depression affected every industrial nation and Britain was certainly not excluded from the misery. All the traditional industries suffered, and even if the unemployment figures were not so bad in the mills as they were, for example, in ship building, that was largely due to the mills going on short-time working where possible. War put an end to mass unemployment, but peacetime only saw a continuing decline. Where once Britain had exported to the world, now cotton goods were being imported. Britain was in an impossible situation, with other countries now having mills as modern as the best in Britain but with lower wages. India, in particular benefitted, simply because they had the raw material available at home. British cotton was being undersold in all the markets of the world. Even at the end of the nineteenth century, there were some who saw this as inevitable, but not everyone mourned its passing. One writer wrote in *The Doom of the Cotton Trade* (1895) of a coffin being built in the East:

> You know what will be buried in this coffin: all the outraged and murdered past, the little apprentice children, tortured as never Inquisition tortured ... There will be buried four generations of blighted and shortened lives, with unnumbered babes born blasted; therein will be buried all the manifold miseries of today, the accursed destroying drudgery of men, women and children; therein will be buried all the pale, sickly faces, the crooked legs and bent backs, the thin frames, the puny wailing infants generated and born on the cheap.

The days of greatness of the industry that changed the world were over. The pioneering work of the eighteenth century had shown how mechanisation could bring prosperity, but the question remained unanswered – prosperity for whom? The legacy of those years was also to create an industrial world of conflict as often as co-operation. If this was true, how did industrialisation affect the countries that followed Britain's lead?

Manchester unemployed in the 1930s.

The new industry that had grown up in the American South certainly prospered, and the poor whites warmly welcomed the opportunity to escape rural poverty for work in the mills. In the North, however, the early enthusiasm for mill work had faded and discontent with pay and conditions led to several attempts to introduce trades unions into the factories. The spread of unionism in the North was received by mill owners with as much enthusiasm as their counterparts had shown in Britain. One result was that mill owners simply moved their operations down to the more tractable South. Here they were frequently able to divert attention from problems by using the racial issue. Nevertheless, the poor conditions, long hours and low pay – and in all these the South was worse off than the North – eventually bred sufficient discontent for the unions to begin recruitment in the South.

The South has always been a conservative region, which makes it the more surprising that it was the more politically committed, communist-led unions that made the greatest impact at first. The National Textile Workers' Union was able to recruit members in spite of their politics, but their strikes were rarely successful. They followed a pattern that was to become common throughout the 1920s. A strike at the Loray mill in North Carolina in April 1929, for example, was quickly broken. Strikers were evicted from their company houses and they set up camp on the edge of town. On 7 June, the police moved in to clear the camp, shots were exchanged and the police chief killed. Sixteen union leaders were arrested, which did little to calm the situation. There were more violent clashes, and a union worker, 29-year-old May Wiggins, was killed. In the end it all came to nothing. The union leader Fred E. Beal jumped bail and escaped to the Soviet Union. He became disillusioned with the realities of Soviet communism and returned to America in 1938, where he was arrested and jailed.

Moderate unions fared no better and were met with equal violence. In June 1929, workers at the Baldwin Manufacturing Company in Marion, North Carolina, joined the United Textile Workers and a public meeting was called to discuss grievances. As a result, twenty-two trade unionists were sacked and the union promptly demanded reinstatement, a reduction of the working day from twelve hours to ten and the establishment of a grievances committee. The terms were all refused, and a strike was called. It was an ill-ordered affair, but some agreement was reached, and management promised the twenty-two would be re-employed. Work resumed, at which point the president, R.W. Baldwin, refused to honour his part of the bargain. There was to be no reinstatement and on 2 October the night shift left the mill and formed pickets to keep the day shift out. Baldwin called in the sheriff, and an eye witness described what happened next to a reporter of the *New York Times*:

The Sheriff ordered the people to stand back … twice he asked them to stand back. 'I am not going to ask you to stand back any more' the Sheriff threatened as he called to workers who wanted to come through the gates 'Come on in, those of you who want to go to work.'

Only one man passed. The Sheriff then jerked out his tear-gas pistol and fired it at Slick Mills, one of the strikers. Mils was blinded and staggered out of the crowd. Deputy Sheriffs Broad Robbins, Charles Tate and James Own the began firing their revolvers.

The crowd scattered, running in all directions. None of our people had guns and none of us tried to resist the officers.

Six workers were killed and twenty-five seriously injured. Baldwin was among the first to applaud the Sheriff's action and all murder charges were dropped. Such cases were by no means uncommon, though none ended as bloodily as that at Marion. Trade unionism never did get a strong footing in Southern mills, though time was to lead to a steady improvement in wages and conditions. The industry had the great advantage of, like India, having a home-grown raw material to use, reducing transport costs to a minimum.

The share-cropping system used throughout the South did little to encourage an intelligent use of the land. Cotton crop followed cotton crop on a land that was becoming increasingly impoverished. At the same time, the crops were attacked by a previously unknown scourge, the boll weevil. The adult insect chews into the boll to lay its eggs and the damage done by the adult is multiplied by the hungry grubs when they hatch. It came to the South from Mexico in 1894 and began moving North at a rate varying from 40 to 160 miles a year. By the end of the century, the pest has been estimated to have cost the South something approaching $20 billion. Between them, erosion and the boll weevil laid waste great tracts of cotton fields. In retrospect it can be seen to have actually benefitted some. Georgia was a state that suffered badly from the pest, and many farmers turned to an alternative crop. They started growing peanuts. This turned out to be so profitable for some Alabama farmers that they erected a statue to the insect in the main square of the city of Enterprise.

That may have been good news for some in Alabama, but the ravages of the boll weevil and impoverishment of the soil still left many croppers struggling. Things did improve but then the depression sent prices tumbling to an all-time low. The New Deal was designed to solve the problem through the Agricultural Adjustment Agency (AAA). In 1933, the AAA diagnosed the problem as over-production – there was too much cotton and that kept the price low. They began a policy of ploughing up the cotton fields, taking over 10 million acres out of production and paying compensation to the owners. This was fine for the landlords, but little help for the croppers, as very little of the relief money trickled down to them. A survey of some 3,000 croppers in Alabama disclosed that 80 per cent were in debt and had been for more than a year. They were found to be living in conditions that had not improved since the end of the Civil War

The boll weevil statue in Enterprise, Alabama.

Just as the mill workers had done, the croppers began to band together and the Southern Tenant Farmers' Association was formed in Arkansas. It was unique among organisations in the South in that it had both white and black members. Blacks were given positions of authority, and formally addressed as 'Mister'. It enraged many Southerners and if methods employed against mill unions were violent, it was nothing compared to the response to the Croppers' Association. A white organiser was told in the crudest terms, 'We don't need no Gawd-damn Yankee bastard to tell us what to do with our niggers.' Black leaders fared even worse in what was to become a reign of terror in the state. A.B. Brookins, the union chaplain, described what happened to him and many others could tell a similar story. 'They shot up my house with machine-guns, and they made me run away from where I lived at, but they couldn't make me run away from my Union.' Croppers were evicted from their homes, but as soon as the authorities had gone, their neighbours took the furniture back in and defended the property. It was during this period that the protest song *We shall not be moved* was first heard.

It was neither union activity nor New Deal legislation that was ultimately to change the life of the croppers. There was a movement in the North in the early years of the twentieth century as new industries, such as automobile manufacture, provided new opportunities for workers. The biggest change, however, came with mechanisation. The mule was replaced by the tractor, new machines took over the work of ploughing, sowing and hoeing. But the most important change came when the machines also took over the work of picking.

In the 1930s, two brothers John D. and Mack Rust, produced the first spindle cotton picker, in which rotating vertical spindles, studded with spikes, pulled the cotton boll off the plant. Their first effort of 1931 was a machine that could pick one bale a day – by 1933, the improved version could pick five. Plantation records from the previous century suggest that at the height of the season when the plants were covered with bolls, the slaves were averaging around 150lbs a day, but that fell to around 120lbs as the season drew to a close. The machine could do the work of thirteen hand pickers. The Rust brothers were genuinely concerned about the effects of their invention. They saw the obvious advantage of taking cotton picking away from human hands, relieving men and women of the back breaking work in the heat of the fields, but they were also worried about creating mass unemployment. They tried at first to limit the use of the machine and then proposed putting profits into a foundation that would fund the displaced farmers. These schemes did little to help and in the 1940s they abandoned production. But machine use continued to grow. In 1946 there just over 100 spindle pickers at work; by 1953 there were over 15,000. Picking was not the only area to show improvements. New cotton gins were introduced which operated a continuous process, in which the raw material was fed in at one end and emerged at the far end already strapped as standard bales.

Hand picking disappeared and the croppers went too. The big plantations returned, owned by individuals but also large corporations. One example of the change is the Delta and Pine Land Company plantation in Mississippi. It was formed out of small plantations in 1911 specifically to serve the British manufacturer, Fine Spinning of Lancashire, that was later taken over by Courtaulds. A company town was formed, and the plantation became a self-contained unit, which at the height of its prosperity was farming 34,000 acres planted with cotton. It was the first to use arsenic to control the boll weevil and the first to use aircraft for crop spraying. The sprayers did so well that they were able to expand their business and it would eventually become Delta Airlines. Delta began breeding better cotton and developed a successful seed business. Courtaulds left in 1978, and the plantation began to diversify – largely moving into soya beans, with some rice and wheat. It is a pattern that has been repeated in much of the traditional cotton growing regions.

Problems developed in the latter part of the twentieth century, when overproduction became a problem. In the year 1982–3, the crop was estimated at 12 million bales, of which 5.4 million went to the domestic market, 5 million to export and leaving 1.6 million bales unsold. The answer was once again to cut production. The government introduced a Payment in Kind (PIK) programme in which farmers were in effect paid for not growing cotton – and not using the land. The government passed on the equivalent in cotton to that which would have been grown on the land. Production was reduced, the farmer got compensation and someone else could worry about what to do with the cotton stocks. The South had changed dramatically since the Civil War and, thanks to the heroic efforts of the Civil Rights movement, segregation has been removed from the statute books, though changing attitudes is always more difficult than passing legislation.

In the American South it is easy to see the influence of the past on everyday life. The same could be said of India, where, if anything, change has been even more dramatic than in either Britain or America. Everything that has happened in Indian cotton in the twentieth century has been dominated by the fight for independence and the social policies adopted since the end of British rule. Indians had long ago recognised that to the British their country existed as a supplier of raw materials and as a market for manufacturers. It was part of an economic empire and could be attacked through the economy. This was the origin of the Swadeshi movement that began in 1905 and encouraged the boycott of British goods. It provided an enormous stimulus for India's mills and an even greater stimulus came through the extended boycott led by Mahatma Gandhi. But Gandhi was not content with the notion that British manufacture should be replaced by Indian manufacture. He wanted to spread the movement so that the effects should not just benefit the minority who lived in towns and worked in factories but also the multitudes of village India.

From Swadeshi there came the Khadi movement, the production of cloth in the villages by traditional means, by spinning on the charkha and weaving on the handloom. Gandhi's views have sometimes been expressed as rather like those of William Morris, little more than a nostalgic yearning for a past age – and Gandhi's

own view of spinning as an almost mystic activity seems to support that view. But he saw Khadi quite clearly as an economic necessity in a country with such a huge problem of rural poverty. He wrote, 'Khadi is the only true economic proposition in terms of the millions of villagers, until such time, if ever, when a better system of supplying work and an adequate wage for every

Mahatma Gandhi spinning in his cabin on his way to the London Conference on India in 1931.

able-bodied person is found in every one of the villages of India.' He saw Khadi as being a movement by the poor for the poor:

> The hand looms have suffered because of their having to sell their output to the same middle class which buys its clothing from foreign and Indian mills ... the problem of rural uplift is to be viewed as one of setting up within each locality numerous lines of production which tends to be locally consumed.

In the 1940s, there was much concern among many Indians that whatever might be the future of Khadi and the cottage industry movement, the process of industrialisation should not follow the path of the West. The National Planning Committee put it in plain terms: 'social injustice, economic discontent, class conflict ... these are the unhappy brood of individualist Industrialism.' In fact, the Indian cotton industry was to follow a very different route from that established in Europe, and still find its own solutions to the problems of industrialisation.

From the first, the Indian mills had to come to terms with the complexities of Indian social life, including the special problem of caste. Unlike British and American mills that simply hired hands directly, Indian mills employed gangs of workers hired by jobbers. The Bombay jobber stood in relation to his gang much as the head man to his village community. He hired the workforce, took

Farmers still use bullock carts to being their cotton to the Karjan co-operative gin.

responsibility for their welfare, arranging credit and finding lodgings – and received 'dasturi' which might well be described as bribes, from those who wished to join the gang. He administered a complex system and was the link between the English or Gujarati-speaking owners and overseers and the largely Maharastra-speaking workers. There were further complications in that the workers still regarded themselves as villagers first and mill hands second. They not only sent money back to the villages, but at harvest times or when a family crisis demanded it, returned there to help out. So, every gang consisted of the 'permanent' hands and 'badlis' or reserves who could be called in to replace the absentees. It seemed to European eyes to be chaotic but, like so many Indian institutions, it fitted the particular circumstance of Indian life very well, but was a half-way stage, a bridge between the life of the village and the industrial town.

As the industry grew, so the jobbing system declined and ultimately disappeared and it seemed that India was following the Western pattern of confrontation between management and workforce. To some extent this did happen, especially in Bombay; elsewhere the system threw up some uniquely Indian solutions. The most remarkable example is the story of the Ahmedabad strike of 1918.

It was a complex dispute that began over payments made to workers who stayed on in the mill when the town was afflicted by plague. The extra payments were later withdrawn and complaints about that became entangled with more general concerns about pay and conditions. When the strike began there were three principal figures involved – Ambalal and Anasuya Sarabhai and Gandhi. The Sarabhai family were leading mill owners, but Anasuya had spent some time in Britain where she was involved in the suffragette movement and when she returned to India, she went to see Gandhi to offer her help. Gandhi pointed out that as she had been brought up in the textile industry, she could best help by organising the workers. The strike began with brother and sister opposing each other. It was more than a strike about money, it was an argument about principle, as Ambalal was clearly aware when he criticised Gandhi's philosophy:

> He assumes that mills are run out of love for humanity and as a matter of philanthropy, that their aim is to raise the condition of the workers to the same level as that of the employers. We beg to say that his approach in this respect is wrong … employment of labour and conditions of employment are determined purely on the basis of supply and demand.

The workers had few resources to sustain them through the strike and Gandhi, to demonstrate his total support, declared that he would not eat while the strike lasted. It was the first of his fasts and had an immediate effect. Ambalal at once proposed that the issue be settled by arbitration while suggesting to Gandhi that the fast was a form of blackmail when used against a friend. Gandhi agreed at once, the strike was over, and he told the mill workers: 'I have never come across the like of it. I have had experience of many such conflicts or heard of them but have not known any in which there was so little ill will and bitterness,'

The end result was the formation of the Ahmedabad Textile Labour Association under the leadership of Anasuya Sarabhai, dedicated to settlement by arbitration rather than by strikes. It has been successful, with Ahmedabad enjoying industrial peace unknown in other parts of India – or indeed in many other parts of the world. It is generally regarded as a satisfactory system, though not without its problems, at least for the owners. As an Ahmedabad mill owner told me, the union now has one of the best teams of experts in industrial relations in India and they rarely lose a case – so rarely in fact that most mill owners settle without waiting for arbitration to save on legal costs.

The most profound change came with independence. First the agonies of partition resulted in the cutting off of many of the principal growing areas from the manufacturing centres, the latter staying in India, much of the former in Pakistan. Indian growers have had to work hard, experimenting with new hybrid plants and new techniques, in order to make India self-sufficient in cotton and even an exporter. The problem lay mainly with the method of cultivation depending on village smallholdings. This was true as late as 1980, when a report showed that this was true even in Gujarat, which produced over a quarter of the country's cotton yet had far less than a quarter of the total cotton acreage. Much of the improvement is due to the co-operative movement working with government backing. A typical co-operative, such as the Karajan Co-operative Ginning and Pressing Society, runs the local gin and press and copes with most of the marketing for local farmers. The farmer brings in his cotton and is immediately paid 75 per cent of the estimated price it will fetch and receives the rest when the crop is sold, after deduction of expenses. The farmer makes on average 10 per cent more than he would have got had he sold it himself at auction.

The co-operative is more than just a ginning and marketing body, it is also a channel for funds and new ideas. Under it are eighty-nine village co-operatives, representing over 4,000 shareholders within a 20-mile radius of the gin. Improvements in crop types, in pest control and cultivation methods are fed down to the level of the smallest village. The result was a gradual, if agonisingly slow, increase in productivity and incomes. No one viewing the co-operative gin in the 1980s with its 100-year-old machinery could fail to see that there is still a long way to go. Statistics for that period are revealing of just how big the gap was between Indian production and American. India as a whole had 19.4 million acres under cotton as compared to 12.8 million in the United States yet produced only 7.8 million bales as against the American production of 15.7 million. American yields were three times that of Indian. Nevertheless, India had shown remarkable improvements in the post-independence years.

The most remarkable feature of all appears when one turns to the production of cotton cloth. At the time of independence, the country was already showing

The Karjan hydraulic press turns out a cotton bale.

a bewildering complexity. There was a large mill sector, Khadi was already established as a major force in the villages and there was a multiplicity of small units, using both handlooms and power looms. It would have been expected that the small units would disappear as the mills grew in number – which they did, from 378 in 1951 to 723 in 1982. More and more yarn was produced by the Indian mills to the point that the industry had returned to the position it had held before the Europeans arrived – growing and spinning all the cotton the country needed and having enough left over to export. The figures show that as expected the number of spindles increased in the same period from 11 million to 21.8 million. But turn to the figures for looms in the mills and things look

The unchanging scene; women wind on bobbins for the handloom weavers of Panipat.

very different – throughout the period the number remained constant at roughly 200,000. There was, it is true, a move towards replacing old looms with modern automatic looms, but the actual numbers scarcely changed. This was not because the owners did not want to install more looms, but they were prevented from doing so. Government had decreed that the wholesale use of labour-saving machinery would make little or no sense if the sole end product was profit for a few and unemployment for thousands. So it was decreed that the cloth woven in India would be divided between three sections – mills, small groups using power looms and handlooms. No one can pretend that the handloom is anywhere near as efficient as the power loom, nor that the power loom can match the modern machinery of the mills. It was a decision entirely based on the needs of vast proportion of the population. And every extra rupee earned by a village worker that lifts him above subsistence level is a rupee he can spend as a customer. To a western visitor who is used to thinking of the handloom as an antique machine of some interest to the arts and crafts part of the textile world, it is a revelation to visit an Indian town and find 40,000 of them at work. It is almost as startling to find a policy that says success cannot be measured by productivity alone. Visiting India in the 1980s, one was inevitably aware of the levels of poverty in the country, but also aware of a certain optimism.

Pride of the co-operative movement: a modern spinning mill at Ichalkaranji, complete with rose bushes.

In Ichalkaranji, in Kolhapur, I visited three spinning mills run by co-operatives formed by the local power loom weavers. They had produced a modern, efficient industry equipped with the latest machinery. There was pride in the mills – and that extended to producing prize roses from the bushes planted outside by the workers. Here, at least, the old pattern of poverty was beginning to crumble. Not so long ago, this was an area like many another, where the Maharajah's palace sat in the middle of cotton fields. The old palace still stands, ornate and beautiful but no longer home to the Maharajah – it has been bought by the co-operative and turned into a training centre for textile students. It stands as a symbol for the changes that swept through India – an India that has once gained a prominent role in its own right in the world's cotton industry.

Chapter 18

EPILOGUE

When I wrote this book I speculated about the future of the industry, especially in India. Forty years later, there have been many changes. One of the most significant is in the basic production of raw material, where the country is now the biggest grower in the world, with an output of 36 million bales a year. The industry is absolutely vital to the whole economy, employing over 50 million people in growing and processing cotton. There have also been changes to the balance in the spinning and weaving sectors. The co-operative movement has not spread significantly, representing only a modest 12 per cent of the mills. What has remained constant is the concentration on spinning in the mills in both the private and the public sector. Three quarters have concentrated purely on producing yarn, and the composite mills that both spin and weave are struggling financially. By contrast, the power loom sector thrives and the current estimate is that there are some 380,000 of the machines in use at workshops with ten or fewer installed. If that seems surprising in the twenty-first century, what is even more surprising is to find the continuing success of the handloom sector where there are a staggering 4 million in use. It is obviously not about efficiency; the power loom sector with only one fifth of the number of looms as the handloom sector still produces nearly two thirds of the cloth. Only 1 per cent of the looms in use in the country are the modern shuttleless looms in use in other parts of the world. It is easy to see why, given the vast numbers employed in the handloom sector, that there is little incentive to engage in a full-scale modernisation programme. But one cannot help wondering how long this can last. India is changing, having already become a formidable economic force in the world. Change will inevitably come; it is only a question of when.

In America, cotton growing is still thriving and where India consumes most of the cotton it produces itself, America is the world's largest exporter with 35 per cent of the world trade, 20 million bales per annum. A great deal of this comes from a successful programme for eradicating the boll weevil. Originally, this depended on a huge use of pesticides, spraying their crops with a mixture of chemicals up to fifteen times a season. It accounted for almost half of the insecticide use for the whole of the United States. The programme of eradication began with a study of the pest's life cycle. It was discovered that treatment with methyl parathion in the autumn reduced the winter stock substantially. In 1978, The United States Department of Agriculture set up an eradication programme in which the Animal and Plant Health Protection Service put up 30 per cent of the costs, with the cotton producers paying the rest. It used biological as well as

A modern cotton picker work on one of the vast cotton plantations of Texas.

chemical methods, and today the boll weevil has almost completely disappeared from the South. The story is told dramatically by the statistics of production from Georgia. In 1915, the state was producing 2.8 million bales of cotton a year. Ten years later, thanks to the boll weevil, that was down to 600,000 and by 1983, it reached a low point of just 112,000 bales. Then the eradication programme came in and by 2000 production had rocketed to 1.66 million bales. There was also good news for the environment: insecticide use had fallen from fifteen times a year to just two. What did come as a surprise when looking at the statistics was to find that by far the largest quantity was not grown in the traditional states such as Mississippi and Georgia, but in Texas. If the growing industry is thriving, the same cannot be said of the mills. They face a very familiar problem, competing with countries such as China, where wages are comparatively low. It is the same competition that has almost eliminated the cotton mills of Britain and would have done so completely if a new company, English Fine Cottons had not taken over a Victorian spinning mill at Dukenfield near Manchester and installed the latest computerised machinery. The secret of their success lies in the fact that they use only the finest long staple cotton to produce a luxury product. They have acquired their own niche in a crowded market, but whether cotton spinning will ever return in any great quantity to Britain is highly dubious. King Cotton abdicated from the British throne years ago and is unlikely ever to regain his crown.

APPENDIX

Life in the mills of Britain and the plantations of America were governed by carefully formulated rules. The rules quoted here for the plantation overseer are typical of many such sets of regulations, whereas the plantation manual is more detailed than most and has been included to show the extent to which an owner would attempt to regulate all aspects of slave life. Mill rules vary in detail, but all seem to have one thing in common: the owner has stretched his imagination as far as possible in attempting to foresee the misdemeanours into which his employees might be drawn.

Rules for the Overseer or Manager, Willis P. Bocock Plantation, Marengo County, Alabama

1. He must not indulge in swearing, drinking, or any immorality on the plantation. The morals of the negroes must be strictly attended to, and unless the overseer conducts himself like a gentleman no improvement can be expected of the negroes.
2. He should not enter into conversation with the negroes except on business. Familiarity breeds contempt.
3. Occasionally and at odd times he should patrol the negro quarters to see nothing amiss is going on.
4. He must not injure the negroes, never strike one in a passion, or with anything that might do mischief. If they do wrong & need correcting let it be done without cruelty! And avoid taunting them with their misconduct afterwards. Avoid torture of mind and body.
5. A brisk lively motion for both Mules and negroes should generally be required, but no rushing. It is desirable for everything to last as long as possible, which cannot be the case if overworked.
6. Breeding & suckling women should be worked near the house if possible, & carefully.
7. Mules & horses should be well rubbed & attended to every day, & negroes should be allowed a specific time for that, say one hour at twelve o'clock in winter, and the times lengthened as the days grow long & hot. In hot weather hands & mules must have 2 hours rest at 12 o'clock.

8. At night let the negroes employ themselves as they please till the bell rings without any interference unless they are violating some rules of the plantation.

9. The women must be allowed time every week for washing and mending, & the largest girls must help their mothers. If a rainy day comes in the week let them have it then. If not, allow two or three hours by Sun on Saturday evenings.

10. About once a month, say on the 1st Saturday, after the weather begins to get warm, have a cleaning up about the houses, yards, and under the houses, and haul off the litter. Cleanliness is necessary to health.

11. Negroes are allowed to come up and see Mistress and Master on Sunday mornings and late Sunday afternoons. At all other times they must ask leave of the overseer before they come. The house hands do not go to the quarters without permission.

12. Men are allowed a rooster and three hens, two hens additional for their wives, and one for each child that works in the field. These they have to start with at the first of the year. The chickens & eggs they raise are for their own use, and not to be sold off the place.

13. As we try to raise Meat and all the supplies for the place, the overseer must pay as much attention to the stock of all kinds as to the crops. Have them counted frequently, & count himself, & list at least once a month.

14. Notice houses fences & gates, put and keep them in as good condition as practicable. Attend to the gear of all kinds, and all plantation tools wagons plows etc; let all be kept in good order & good repair, and when not in use put in their right place, not exposed to hot sun and wet weather.

15. The great rule of managing is – Keep everything in good condition & repair. Waste nothing; take care of everything.

16. Especially when likely to be scarce. And the rule for cropping is, first make an abundance to live and feed on, including plenty of vegetables for the negroes, then as much market crop as convenient.

17. The overseer raises nothing for sale, except for his employer, and sells nothing off the place except by authority of his employer: then for cash only which he at once hands over.

18. The overseers' time is paid for by his employer. It is not right for the overseer to use that or anything else that belongs to his employer, in going about, in visiting, or in any way but for his employer.

19. He keeps a book, prepared & furnished him at the beginning of each year, in which the rules are written down, & he sets down every thing in the course of the year according to the heads.

20. And when his employer is absent he writes to his employer at the end of every week an account of matters on the place. Sending at the end of every month a report of his stock &c.

Marengo County, Alabama, July 15th 1860

This agreement witnesseth that Lewis A. Collins has agreed to live for Willis P. Bocock at Waldwick as his supervisor or manager from the above date till the end of this year. He agrees to give his whole time & attention, his best skill, industry & judgement, to carrying on the business and carefully managing every thing under his charge; to attend to and observe the wishes & directions of his employer when made known; to be careful of the good conduct health & cleanliness of the negroes; to take care of the sick; & to see they comply in a reasonable manner with the rules laid down before them; also to conduct himself with prudence & sobriety, and a faithful regard to his employers' interest.

Said Bocock on his part in consideration of said Collins' service as rendered, to furnish him with an animal to ride about said business, he furnishing his own saddle and bridle; to allow his family that stay with him on the place a reasonable finding of such things as are raised on the plantation to be used without waste, also a woman of the place to cook & wash for them, and to pay his wages at the end of his time at the rate of five hundred dollars for twelve months, said Collins accounting for all time lost from his employers' business.

As evidence that said Collins intends to be faithful it is further agreed that whenever said Bocock becomes dissatisfied he may put an end to this agreement by paying sd Collins up to that time; and further at the said Collins own request, that if he drinks any spirits in the time he is not to receive any wages.

July 16, 1860 Lewis O. Collins
W.P. Bocock

Plantation Manual, origin unknown

Allowances

Allowances are given weekly. No distinction is made among work-hands; whether full hands or less than full hands, or adjuncts about the yard, stables, etc. A peck of meal apiece is given every Sunday morning. The peck measure is filled and piled up as long as will remain in it, but not packed or shaken. Meat and syrup are given out on Monday night. When meat alone is given, 3 pounds of pickled pork, or bacon, apiece is the allowance. Fresh meat, salted over night, may be given at the rate of 3½ pounds of beef or pork. In summer but one-half of the allowance may be fresh meat. As soon as cold weather sets in fresh pork allowance begins. Of hog offal 4 pounds. With one quart of syrup, the meat allowance is reduced 1¾ pounds of pickled pork, or 2½ of hog offal. No deductions are made for light sickness of a day or so, or for pregnancy. A ditcher who does each task without occasioning annoyance for a week receives on Wednesday night an extra pound of meat. The driver is allowed a small extra of meat and molasses whenever he may apply for it, which is but rarely done. Each ditcher receives every night, when ditching (fall & winter) a dram (jigger) consisting of 1/5 water and 4/5 whiskey, with as much asafoetida as will absorb, and a long string of peppers in the barrel. The dram is a good sized wine glass full. In cotton picking times, when sickness is prevalent, every hand gets a dram before leaving for the field. After a soaking rain all exposed to it will get a dram before changing their clothes. Drams are never given as rewards and only as medicinal. From second hoeing, or early in May an occasional allowance of tobacco is given to those who use it, about 1/8 of a pound, usually after some general operation as a hoeing, ploughing &c. This is continued until the crops are gathered when each negro can provide for himself. Each *man* gets in the fall 1 cotton shirt and 1 pair woollen pants and 1 woollen jacket. In the spring they get 1 shirt and 1 pair of cotton pants. Each *woman* gets six yards of woollen cloth and 3 yards of cotton shirting in the fall, with a needle, skein of thread and ½ dozen buttons. In the spring they get 6 yards of cotton drillings, 3 yards of shirting with needle thread and buttons. A stout pair of shoes is given to each in the fall and blanket every third year.

Children
There is a separate apartment under the charge of a trusty nurse where the children are kept during the day. Weaned children are brought to it at the last horn blown in the morning, about good day light. The unweaned are brought in at sun rise after suckling, and left in charge of the nurse. Allowance is given out daily to the children. 4 quarts of meal, 3 quarts of hominy and 2 pounds of bacon are found sufficient for 20 children. They also have 1 pint of skimmed milk each day. Their breakfast consists of hominy and milk. At mid-day their meat is made into soup-pot liquor generally, with vegetables to be boiled in it if any are to be had and dumplings or bread. They are never allowed fresh meat except the bony parts of beef to make soup. 1½ pints of molasses are given among the same number every Wednesday morning. Each child gets a shirt and the girls a frock also; the boys a pair pantaloons reaching the neck and with sleeves every fall and spring, of the same goods as the work hands. A child's blanket is given to them every third year. Children born can have a blanket at the time of birth or the fall following according to the necessities of the Mother. All children are required to appear in entirely clean clothes twice a week. It is the duty of their mothers to attend to it and the nurse to see that it is done or report it. The nurse must also see that no child changes its clothes from thicker to thinner clothes or the reverse at improper times or has on any wet garment at any time.

Plantation hours
The first morning horn is blown one hour before breakfast. Work hands are expected to rise and prepare the cooking etc. for the day. The second horn is blown at good daylight, when it is the duty of the driver to visit every house and see that al have left for the field. The plough hands leave their houses for the stables at the summons of the plough driver 15 minutes earlier than the gang,

the overseer opening the stables. At 11½ A.M. the plough hands stop to feed. At 12 A.M. the gang stop to eat dinner. At 1 P.M. through the greater part of the year all hands return to work. In summer the intermission increases with the heat to 2½ hours. At 15 minutes before sunset the plows stop for the day and at sun set the rest of the hands. No nightwork is ever exacted. At night the negroes are allowed to visit among themselves until horn-blow – at 8½ o'clock in winter and 9 o'clock in the summer, after which no negro must be seen out of his house and it is the duty of the driver to go around and see that he is in it.

Sucklers are not required to leave their houses until sun-rise when they leave their infants at the children's house before going to the field. The period of suckling is one year. For six months they return three times a day to suckle their infants – in the middle of the morning, at mid-day and at the middle of the afternoon. Their work lies always within a quarter of a mile of the quarter. They are required to cool before commencing to suckle, to wait fifteen minutes at least in summer after reaching the children's house before nursing. It is the duty of the nurse to see that none of them are heated when nursing as well as of the overseer and his wife occasionally to do so. They are allowed 15 minutes at each nursing to be with their children. At 7 months they return to nurse but twice a day, missing out mid-day. Each woman on weaning her child is required to put it in charge of some woman without a child for two weeks and not to nurse it at all during that time. A suckler does about ⅗ of a full hands work, a little increased towards the last.

Such hands as cannot follow with the prime gang from age or deformity are put with the sucklers. Pregnant women at 5 months are put with the sucklers gang.

The regular plantation midwife shall attend all women in confinement. Some other women learning the art usually assist her. The confined woman lies up one month, the midwife remaining with her the first 7 days.

There is a hospital adjoining the children's house where all the sick are confined. Every reasonable complaint is promptly attended to and with any marked or general symptom of sickness a negro may lie a day or so at least. Homeopathy is exclusively used. As no physician is allowed to practice on the place, there being no homeopathist convenient, each case has to be examined carefully by the master or overseer to ascertain the disease. The remedies next are to be chosen with the utmost discrimination. The vehicles for preparing and administering with are to be thoroughly cleansed. The directions for treatment, diet, etc. most implicitly followed; The effects and changes cautiously observed, and finally the medicines securely laid away from accidents and contaminating influences. In cases where there is uncertainty the books must be taken to the bedside and a thorough and careful examination of the case and comparison of medicines made before administering them. The head driver is the most important negro on the plantation. He is to be treated with more respect by both master and overseer. He is on no occasion to be treated with any indignation calculated to lose the respect of the other negroes without breaking him. He is required to maintain proper discipline at all times. To see that no negro idles or does bad work in the field, and to punish it with discretion on the spot. The driver is not to be flogged except by the master but in emergencies that will not admit of delay. He is permitted to visit the master at any time without being required to get a card though in general he must inform the overseer when he leaves the place and presents himself on returning. He is expected to communicate freely whatever attracts his attention or he thinks information to the owner.

Marriage is to be encouraged as it adds to the comfort, happiness and health of those entering upon it, beside assuring a greater increase. No negro can have a wife, nor woman a husband, not belonging to the master. Where sufficient cause can be shown on either side a marriage may be broken but the offending party may be punished. Offenders cannot marry again after such divorces for three years. As an encouragement to marriage the first time any two get married a bounty of $5.00 to be invested in household goods or an equivalent of articles shall be given. If either has ever been married before the bounty shall be $2.50 or equivalent. A third marriage shall not be allowed but in extreme cases or where both have been married before no bounty shall be given No marriage shall take place without the master's express consent to it.

Negroes living at one quarter having wives at the other, are privileged to visit them only between Saturday night and Monday morning and must get a pass for each visit. The pass card must be delivered immediately on reaching the destination and a return card given ready to return. All are subject to the regulation of the place they are at any time upon and it is as much the duty of the overseer and driver to observe them as the others under their ordinary charge. Each male work-hand shall be allowed to go to town once a year, on a Sunday between crop-gathering and Christmas. Not more than ten can go the same day. Adjoining each negro house is a piece of ground convenient for a garden. They also have patches to cultivate little crops of their own. There is also a small field planted and worked generally in pindars [ground nuts] the same as the rest of the crop the produce of which is divided among them. At Christmas there or four days holyday are given on one of which is a barbecue of beef or mutton and pork, bread and coffee are provided. Also a holyday and cue in August.

Horses and Mules

Every horse and mule when in use is curried in the morning before taken from the stable and at 12 o'clock; also at night whenever they appear unusually sweaty and fatigued. They are fed twice a day; at night as much corn as they can eat, generally nearly a peck and about 8lbs. of fodder, hay or shucks apiece: and at 12 o'clock 6 ears of corn and about 2½ of fodder each. When idle they require one third less and are curried but once a week, on Sundays. Watering is done three times a day when idle and 4 times a day when at work: always morning and night and before and after 12 o'clock feeding when at work but only after when idle. One gill of salt of best quality is given every other day unless soaked corn is used. When soaked corn is fed it is put in a barrel in the morning and ⅔ of a gill of salt to each mule sprinkled over it and water enough added to cover it. This fed only at night. A gill of dry salt besides is given to each horse twice a week. Horses are haltered in separate stalls. Mules are left loose in another stable. Fresh straw is added daily so as to keep a dry bed always at night.

Hogs in pasture are fed every other day. They are not to be fed constantly at one spot. The amount given at a feeding is about 1 bushel of inferior corn to every 50 little and big. The overseer attends to this in person assisted by such negroes as may be required. A counting of them must be made and entered on the plantation book at least once every month and marking, cutting and speying at least once in every three months. No hogs must be left at large except a few shoals about the lot.

Rules for Water-Foot Mill, Haslingden, September 1851

1. All the overseers shall be on the premises first and last.
2. Any person coming too late shall be fined as follows:- for 5 minutes 2d, 10 minutes 4d and 15 minutes 6d, &c
3. For any Bobbins found on the floor 1d for each bobbin
4. For single Drawing, Slubbing, or Roving 2d for each single end.
5. For Waste on the floor 2d.
6. For any Oil wasted or spilled on the floor 2d each offence, besides paying for the value of the Oil.
7. For any broken Bobbins, they shall be paid for according to their value, and if there is any difficulty in ascertaining the guilty party, the same shall be paid for by the whole using such Bobbins.
8. Any person neglecting to Oil at the proper times shall be fined 2d.
9. Any person leaving their Work and found Talking with any of the other workpeople shall be fined 2d for each offence.
10. For every Oath or insolent language, 3d for the first offence, and if repeated they shall be dismissed.
11. The machinery shall be swept and cleaned down every meal time.
12. All persons in our employ shall serve Four Weeks' Notice before leaving their employ; but L. Whitaker & Sons, shall and will turn any person off without any notice being given.
13. If two persons are known to be in one Necessary together they shall be fined 3d each; and if any man or boy go into the Women's' Necessary he shall be instantly dismissed.
14. Any person wilfully or negligently breaking the Machinery, damaging the Brushes, making too much Waste &c., they shall pay for the same at its full value.
15. Any person hanging anything on the Gas Pendants will be fined 2d.

16. The Masters would recommend that all their Workpeople wash themselves every morning, but they shall wash themselves at least twice every week, Monday Morning and Thursday morning, and any found not washed will be fined 3d for each offence.
17. The Grinders, Drawers and Rovers shall sweep at least eight times in the day as follows, in the Morning at 7½, 9½, 11 and 12; and in the Afternoon at 1½, 2½, 3½, 4½ and 5½ o'clock; and to notice the Board hung up, when the black side is turned that it is time to sweep, and only one quarter of an hour is allowed for sweeping. The Spinners shall sweep as follows, in the Morning at 7½, 10 and 12; in the afternoon at 3 and 5½ o'clock. Any neglecting to sweep at the time will be fined 2d for each offence.
18. Any person found Smoking on the premises will be instantly dismissed.
19. Any person found away from their usual place of work, except for necessary purposes, or Talking with any out of their own Alley will be fined 2d. for each offence.
20. Any person bringing dirty Bobbins will be fined 1d for each Bobbin.
21. Any person wilfully damaging this Notice will be dismissed.

The Overseers are strictly enjoined to attend to these Rules, and they will be responsible to the Masters for the Workpeople obeying them.

BIBLIOGRAPHICAL SOURCES

Documentary Sources
Unless otherwise stated, the quotations relating to plantations and slavery are from Southern Historical Collection, University of North Carolina; information on Styal is from the Greg papers, Manchester Reference Library; the Oldknow Papers, John Rylands Library, Manchester.

Published Works
Baines, Edward, *History of the Cotton Manufacture of Great Britain*, 1835
De Bow, J.D.B., *The Industrial Resources of the Southern and Western States*, 1852
Ellison, Thomas, *The Cotton Trade of Great Britain*, 1886
English, W., *The Textile Industry*, 1969
French, Gilbert J., *Life and Times of Samuel Crompton*, 1859
Fritton, R.F. and Wadsworth, A.P., *The Strutts and the Arkwrights 1758-1830*, 1958
Gadgil, D.R., *The industrial Evolution of India in Recent Times*, (5th ed.) 1971
Roll, Eugene d. Genovese, *Jordan Roll*, 1975
Lord, John, *Memoir of John Kay*, 1903
Green, Constance Mclaughlin, *Eli Whitney and the Birth of American Technology*, 1956
Owens, Leslie Howard, *This Species of Property*, 1976
Mehta, S.D., *The Cotton Mills of India*, 1954
Strickland, M., *A Memoir of Edward Cartwright DD*, 1843
Van Woodward, C., *Origins of the New South*, 1951
Ure, Andrew, *The Philosophy of Manufacture*, 1835

PICTURE CREDITS

INDEX